D0204878

HUMAN–LIVESTOCK INTERACTIONS

The Stockperson and the Productivity and Welfare of Intensively Farmed Animals

HUMAN–LIVESTOCK INTERACTIONS
The Stockperson and the Productivity and Welfare of Intensively Farmed Animals

Paul H. Hemsworth

Animal Welfare Centre, Agriculture Victoria, Australia and Institute of Land and Food Resources, University of Melbourne, Australia

and

Grahame J. Coleman

Animal Welfare Centre, Department of Psychology, Monash University, Caulfield, Victoria, Australia

CAB INTERNATIONAL

CAB INTERNATIONAL
Wallingford
Oxon OX10 8DE
UK

Tel: +44 (0)1491 832111
Fax: +44 (0)1491 833508
E-mail: cabi@cabi.org

CAB INTERNATIONAL
198 Madison Avenue
New York, NY 10016-4314
USA

Tel: +1 212 726 6490
Fax: +1 212 686 7993
E-mail: cabi-nao@cabi.org

A catalogue record for this book is available from the British Library, London, UK.

Library of Congress Cataloging-in-Publication Data

Hemsworth, Paul H.
Human–livestock interactions : the stockperson and the productivity and welfare of intensively farmed animals / Paul H. Hemsworth and Grahame J. Coleman.
p. cm.
Includes bibliographical references (p.) and index.
ISBN 0–85199–195–5 (alk. paper)
1. Livestock industry––Moral and ethical aspects. 2. Livestock––Research––Moral and ethical aspects. 3. Animal welfare.
I. Coleman, Grahame J. II. Title.
HV4757.H46 1998
174' .3––dc21 97–43256
CIP

ISBN 0 85199 195 5

Typeset by Wyvern 21 Ltd, Bristol
Printed and bound in the UK by Biddles Ltd, Guildford and King's Lynn

Contents

Preface

In 1979 Gross and Siegel recognized the impact of human–animal interactions on their experimental animals. In concluding on the results of three experiments examining the effects of human contact on poultry, these scientists proposed that 'providing only for their physical needs does not result in superior experimental animals. Also important factors in the outcome of experiments will be gentle care and familiarity of birds with the animal handlers and experimenters.' Although the implications were obvious for experimental animals, the authors did not reflect on such implications for commercial livestock. Since this early comment, a number of writers have explored the role of the relationship between the animal experimenter and the experimental animal on the behaviour and physiology of the animal. A reflection of this interest is the book edited by H. Davis and A.D. Balfour in 1992, entitled *The Inevitable Bond – Examining Scientist–Animal Interactions*, which was devoted to examining the extent and diversity of these effects in animal research (Davis and Balfour, 1992).

The effects of human–animal interactions on livestock appear to have been ignored or neglected until recently. This lack of interest was presumably due to industry personnel and animal scientists considering that either the intensity and frequency of these interactions were low enough to render the effects of any negative interactions ineffective on the animals or the type of interactions were inoffensive to the animals. In the 1970s and 1980s, several authors reported on the practical implications of studies examining human–animal interactions in agriculture. In the early 1970s, M.F. Seabrook (1972a) reported that 'herds where the cows readily approached the herdsman had a significantly higher yield per cow than herds where the cows lacked this confidence'. He concluded that the herdsman's personality influenced cow behaviour and productivity. Similarly, one of the authors of this book and colleagues from the University of Utrecht in the Netherlands reported in 1981 (Hemsworth *et al.,* 1981b) a significant relationship between the approach behaviour of breeding pigs to an experimenter in a standard test and reproductive performance. The authors of this publication concluded that 'the reproductive performance of the farm is associated with the relationship developed between the stockman and his breeding stock'.

Since these early reports a number of studies, both in the laboratory and in the livestock industries, have examined the development and consequences of human–animal relationships in agriculture. However, our progress in understanding these relationships in agriculture has been relatively slow compared with other developments in the field of animal science. The study of stockperson characteristics in livestock production creates a number of problems that are generally not encountered when studying other more traditional areas of livestock production. Stockperson characteristics are not as amenable to study as other factors, such as nutrition, housing, genetics, etc., because of our limited ability to: (i) manipulate individual characteristics; (ii) control others not under direct study; and (iii) study humans in commercial situations. These problems have been exacerbated by the lack of interest shown by psychologists in this important area.

Recently, there has been considerable progress in research on human–animal interactions in agriculture. This research has been multidisciplinary in nature and much of the material presented in this book is a mix of agricultural and animal science and psychology. This presents something of a dilemma when attempting to identify the target audience for this book and, therefore, how to pitch the material so that readers will not find it too technical within a particular discipline but, at the same time, will gain insight into the processes that are being described.

This book is aimed at those people who have an interest in human–animal interactions in agriculture, perhaps as trainers, managers of farms in which animals are intensively farmed, or students and academics seeking an introduction to the subject. We have attempted to make the book as self-contained as possible, by giving a brief account of the theories or principles underlying the research discussed.

The first half of this book contains a detailed review of our empirical knowledge of human–animal interactions and, together with the development of the theory underlying this empirical research, the second half of the book leads into an examination of the opportunities to manipulate these interactions. It is our intention with this book to stimulate interest and exploration of the subject by animal scientists and industry personnel interested in the role of the stockperson in determining animal performance and welfare. The subject is particularly relevant to those with responsibilities in the areas of staff training and selection in the livestock industries. This topic is also relevant to the livestock industries' efforts to attract and retain desirable staff.

As we conclude in the final chapter, this is a relatively new direction for industries in which stockpeople regularly interact with livestock. Much has been done to improve genetics, nutrition, health and housing, but efforts to target the stockperson, who performs such a key function, have just begun. We should not underestimate the role and impact of the stockperson on animal performance and welfare. To do so will put at serious risk the performance and welfare of our livestock. It is likely that both the livestock industries and the general community will place an increasing emphasis on ensuring the competency of stockpeople to manage our livestock; the livestock industries' interests are likely to be motivated by both animal productivity and welfare and the general community's interest in animal welfare.

P.H. Hemsworth and G.J. Coleman

ACKNOWLEDGEMENTS

We would like to thank the many colleagues who have actively collaborated with us in studying this topic and who have also provided important encouragement and stimulation: J.L. Barnett, A. Brand, P.J. Dzuik, H.W. Gonyou, the late A. Hoogerbrugge, R.B. Jones, E.O. Price, N.T.C.J. Salden and P. Willems.

We are also indebted to a number of Australian agricultural funding bodies who have had the foresight and commitment to support our research over the last 20 years, which has contributed substantially to the content of this book: Chicken Meat Research and Development Council (Rural Industry Research and Development Corporation), Dairy Research and Development Corporation, Egg Industry and Research and Development Council (Rural Industry Research and Development Corporation) and Pig Research and Development Corporation. The Victorian State Government has also contributed considerably to this research programme.

The constructive comments and suggestions by J.L. Barnett on drafts of the book are appreciated, as is the support in preparation of the typescript and figures by H. Knowles.

Chapter 1

Introduction: the Stockperson as a Professional – Skills, Knowledge and Status

1.1. INTRODUCTION

Human–animal interactions are a key feature of modern livestock production and research has shown that the quality of the relationship that is developed between stockpeople and their animals can have surprising effects on both the animals and the stockpeople. For example, there is good evidence, based on handling studies and observations in the livestock industries, that human–animal interactions may markedly affect the productivity and welfare of farm animals. By influencing the behavioural response of the animal to humans and, in particular, the ease with which animals can be observed, handled and managed by the stockperson, human–animal interactions may also have implications for a number of work-related characteristics of the stockperson, such as job satisfaction, and thus may have a substantial impact on the stockperson. The first objective of this book is to review human–animal interactions in agriculture and, in doing so, examine their implications for both the animal and the stockperson.

A number of human and animal characteristics influence human–animal interactions, which, in turn, may have marked effects on both partners. Thus an understanding of these key human and animal characteristics and an ability to manipulate at least some of them may offer the livestock industries opportunities to provide benefits for both their animals and their stockpeople, in order to improve industry economics and sustainability. The second objective of this book is to explore the key human characteristics that influence animal behaviour, performance and welfare. If characteristics of the stockperson are important determinants of human–animal interactions, opportunities may exist to improve animal performance and welfare in those situations in which the human–animal relationship is poor. Thus the third objective of the book is to examine the opportunities for industry to improve these key human characteristics through stockperson training and selection. The first half of this book contains a detailed review of our empirical knowledge of human–animal interactions (Chapters 3–5), which, in the second half of the book (Chapters 6–8), leads to the discussion of the theory underlying this empirical research and an examination of the opportunities to manipulate these human–animal interactions.

This first chapter, together with Chapter 2, attempts to set the stage by examining the role of the stockperson in terms of both the stockperson's role in agriculture and the ethical and welfare issues relating to intensive farming of livestock. In particular, this first chapter considers the role of the stockperson, focusing on his or her skills and knowledge that are required to achieve high animal performance and welfare.

1.2. THE ROLE OF STOCKPEOPLE

Any reasonable assessment of the role of stockpeople in modern agriculture indicates that stockpeople are professional managers of animals, integral in determining animal performance and welfare. Yet there appears to be a general lack of appreciation of this by people within agriculture, including stockpeople themselves. It is one of the important contentions of this book that the recognition of stockpeople as key professional managers of livestock is an important cultural change that is required within agriculture; such a change is likely to have implications for the image and self-esteem of stockpeople and the opportunities for training stockpeople, which in turn are likely to be highly influential in affecting animal performance and welfare.

In some sectors of agriculture, stockpeople are recognized as important resources and consequently policies on the development of this human resource have been introduced. Unfortunately, these examples in agriculture are often only evident in some of the large corporate enterprises in which investment, often from outside agriculture, has brought with it sound business management practices, including human-resource management skills. Owner-operators of farms may also undervalue their contribution as livestock managers to animal performance and welfare. There needs to be a widespread recognition and appreciation of the important role of the stockperson in livestock production; such a cultural change will in itself facilitate the appropriate management of this important human resource. For example, appropriate staff selection and training policies and other strategies to select, retain and further develop this key resource are likely to become increasingly widespread as the animal industries recognize the impact of stockpeople on animal performance and welfare, and thus industry profitability and sustainability. These developments in the management of human resources should not necessarily remain in the domain of the large corporate enterprises, since commercial and government services have the opportunity to provide such support for smaller enterprises in the interests of industry economics and sustainability.

One of the most famous pronouncements on the role of the stockperson in agriculture was contained in the *British Codes of Recommendations for the Welfare of Livestock* (Ministry of Agriculture, Fisheries and Food, 1983): 'Stockmanship is a key factor because, no matter how otherwise acceptable a system may be in principle, without competent, diligent stockmanship, the welfare of animals cannot be adequately catered for.' Unfortunately, it is debatable whether these sentiments have been fully accepted by the livestock industries and others.

Few studies have attempted to quantify the contribution of the stockperson to animal productivity and welfare. This is partly due to the difficulty of such research but also because of the industry's focus on technological developments in agriculture. Some of the research documenting the stockperson's contribution will be discussed later in this book, but it is useful at this point to consider a preliminary study at our laboratory which provides an indication of the contribution of the stockperson to the productivity and welfare of farm animals.

A recent preliminary study (Pedersen *et al.*, 1997) has provided limited evidence that positive handling by stockpeople may ameliorate the chronic stress response associated with an aversive housing system. Research on pigs has consistently shown that pregnant sows housed in tether-stalls of a specific design will experience a sustained elevation in the basal concentrations of the stress hormone cortisol, indicative of a chronic stress response (Barnett *et al.*, 1989, 1991). In the study by Pedersen and colleagues (1997), 24 pregnant pigs housed in stalls with neck-tethers were randomly assigned to one of three handling treatments: positive handling in which pigs were patted or stroked on approach; minimal human contact; and negative handling, in which pigs were briefly shocked with a battery-operated goad or prodder whenever they approached. The positive and negative handling treatments were imposed daily for 3 min and the experimenter squatted in front of each pig's stall to impose the appropriate treatment. Daytime profiles of plasma free cortisol, measured in isolation from humans, were significantly lower in the pigs in the positive handling treatment than those in the minimal and negative handling treatments (Table 1.1). These preliminary results demonstrate the importance of human factors for pig welfare and indeed suggest that, in this situation, human factors may be relatively more important to animal welfare and productivity than the housing system *per se*; it remains uncertain if the principle extends to the range of current industry practices. Chapter 3 will discuss other studies that highlight the importance of human factors for farm-animal performance and welfare.

This chapter will consider the factors that need to be addressed when defining the role of the stockperson. The underlying principles drawn from industrial/organizational psychology will be discussed first, followed by an overview of the skills, knowledge and status of stockpeople.

Table 1.1. Cortisol concentrations of 24 pregnant pigs housed on tethers receiving either positive, negative or minimal handling, 3 min per day for 4 weeks (from Pedersen *et al.*, 1997).

	Handling treatment		
	Positive	Negative	Minimal
Daytime mean cortisol concentrations (nmol l^{-1})	2.9	7.4	6.1

1.3. CHARACTERISTICS OF THE STOCKPERSON'S JOB

Traditionally, stockpeople have been regarded as unskilled labourers. The role of the stockperson as the key person responsible for the day-to-day welfare and productivity of the animals under his or her care has not received due acknowledgement. In fact, stockpeople are professional managers of livestock. Farm owners and the general community (through governments) entrust the welfare and performance of large numbers of animals to the care of stockpeople. A duty statement for a modern stockperson may read as follows.

1. A good general knowledge of the nutritional, climatic, social and health requirements of the farm animal.
2. Practical experience in the care and maintenance of the animal.
3. Ability to quickly identify any departures in the behaviour, health or performance of the animal and promptly provide or seek appropriate support to address these departures.
4. Ability to work effectively independently and/or in teams, under general supervision, with daily responsibility for the care and maintenance of large numbers of animals.

The stockperson therefore is required to possess a basic knowledge of both the behaviour of the animal and its nutritional, climatic, housing, health, social and sexual requirements, together with a range of highly developed husbandry and management skills to effectively care for and manage farm animals. For instance, stockpeople may have knowledge and skills in a number of diverse management and husbandry tasks, such as oestrus detection (Fig. 1.1) and mating assistance; semen collection, semen preparation and artificial insemination; pregnancy diagnosis with ultrasonography; artificial rearing of early weaned animals; milk harvesting; controlling and monitoring of feed intake to optimize growth, body composition, milk production and reproductive performance; pasture management to optimize pasture production; routine health checks; monitoring and adjusting climatic conditions in indoor units; administering antibiotics and vaccines; shearing and crutching of sheep; teeth and tail clipping of pigs; castration of males; and effective and safe animal handling. These are highly skilful tasks and stockpeople are required to be competent in many of them. Clearly, the training

Fig.1.1. Stockpeople in modern piggeries are required to detect oestrus in female pigs in order to accurately time either artificial insemination or mating.

of stockpeople to develop these competencies should be a systematically and soundly implemented process, in which the requirements of both the stockperson and the industry are addressed.

The conditions in which stockpeople are required to work may differ within and between animal industries and also from that encountered in non-agricultural industries. Stockpeople are often required to work unconventional and unsocial hours and, at times, under unpleasant conditions. Climatic conditions may vary markedly in extensive livestock production and stockpeople in these industries are required to work outdoors in extreme weather conditions. The standard of workplace amenities for both animals and workers may also vary considerably within each type of production system within this continuum. For example, in some indoor units effluent odours and dust levels may be offensive.

The requirements of the job and workplace conditions are onerous, demanding ones, often not widely and fully recognized by the animal industries and others.

1.4. THE STOCKPERSON: IMAGE AND SELF-ESTEEM

In contrast to the views discussed earlier and contained in the *British Codes of Recommendations for the Welfare of Livestock* (Ministry of Agriculture,

Fisheries and Food, 1983) and the implications of studies such as that by Pedersen and colleagues (1997), surveys on the self-esteem of stockpeople and their perceptions of their image outside agriculture often portray a disturbing picture. For example, Beynon (1991) reported that the families of pig stockpeople regarded employment in the pig industry as having a low status. Industry leaders commonly quote the poor image of stockpeople as a factor contributing to the problem of attracting people to and retaining staff in the agricultural industries. In a compassionate plea for recognition of the role of stockpeople, English *et al.* (1992) suggest that stockpeople are the world's most undervalued profession.

1.5. FITTING THE STOCKPERSON TO THE JOB

Given that it is possible to describe the role of the stockperson in the way we have just done, what characteristics should a stockperson have to do the job effectively? After all, if it is accepted that working as a stockperson is a profession, it is important to identify the personal characteristics which best allow a person to meet the job requirements. These characteristics will be a mix of unlearned factors (traits) and learned factors (skills, knowledge, etc.). These will be discussed in this chapter from a largely theoretical perspective, with a more detailed review of current empirical research in later chapters.

1.5.1. Skills and knowledge

The single most important factor in job performance is the skills that the person brings to the job. Knowing and being skilled at the techniques that must be used to accomplish the task are clearly prerequisites to being able to perform that task. For example, in the pig industry, a stockperson working with the breeding herd must be good at oestrus detection and at assisting matings and, if he or she does not have these skills, production will be severely impaired. The stockperson must also have a good working knowledge of the nutrition, housing, social and health requirements of the livestock. Less obvious, and also the subject of limited research, is the impact of stockperson skills in handling and interacting with intensively and extensively farmed animals. Research has shown that most stockpeople in the pig and dairy industries do not know what aspects of routine handling farm animals find aversive, despite the fact that it has been shown that aversive handling has marked negative consequences for the animal and its productivity and welfare (Hemsworth *et al.*, 1993). This will be the subject of extensive review in Chapter 3.

1.5.2. Personality

In the past, personality was thought to be largely irrelevant for the purposes of predicting how well a person would perform in a job. Studies consistently showed that personality did not contribute to the prediction of job performance and most textbooks report that personality is not a suitable measure for selecting individuals for a job. It is now thought that part of the reason for this is that, until the last 10 years, researchers did not agree on the structure of personality and, therefore, it was very difficult to evaluate research findings. However, there is now general acceptance of a 'big five' theory of personality traits (Barrick and Mount, 1991). Today most researchers agree that personality can be characterized in terms of 'extraversion/introversion', 'emotional stability', 'agreeableness', 'conscientiousness' and 'intellect'. Extraversion is associated with sociability, assertiveness and an outgoing nature. Emotional stability refers to a trait similar to Eysenck's neuroticism and includes such things as anxiety, embarrassment and insecurity. Agreeableness is associated with cooperation, good nature and tolerance. Conscientiousness is characterized by dependability and being hard-working and persevering. Intellect includes being imaginative, cultured and original (Barrick and Mount, 1991).

These personality factors appear to be useful in matching people to some kinds of jobs. For example, a salesperson might be expected to be high on extraversion and a music director high on intellect. In agricultural industries, there is little evidence relating personality to good work performance. Two exceptions are the studies by Seabrook (1972a, b) and Ravel *et al.* (1997). Seabrook (1972a, b) reported that the stockperson's personality was related to behaviour of the cows and milk yield of the herd. The highest-yielding herds were those where the stockpeople were introverted and confident and where the cows were most willing to enter the milking parlour and were less restless in the presence of the stockperson. Significant relationships have been found in the pig industry between personality types of stockpeople and productivity in farrowing units (Ravel *et al.*, 1997). While self-discipline was a trait that appeared to be important at all farms studied, high insecurity and low sensitivity were favourable traits in relation to piglet survival at independent owner-operated farms, while stockpeople that were highly reserved and bold, suspicious, tense and changeable were associated with higher piglet mortality at large integrated farms.

Of considerable possible relevance to productivity is the general finding by Barrick and Mount (1991) that the 'big five' personality traits may be related to worker performance. Conscientiousness appears to be associated with a person accomplishing tasks (see also Barrick *et al.*, 1993). Extraversion is associated with training proficiency. Clearly, to identify the characteristics of a good stockperson and to develop a procedure for selecting a good stockperson, an understanding of the relevance of personality is essential. Any model that attempts to characterize a professional stockperson and to develop a selection procedure for stockpeople should take personality factors into account, particularly because personality may affect other characteristics that are relevant to job performance, such as skills and knowledge. As will be discussed in some detail in Chapter 4, personality may influence the stockperson's attitudes towards animals under his or her care. Tett *et al.* (1991) carried out a meta-analysis in which they systematically evaluated all relevant research on personality measures as predictors of job performance and concluded that, contrary to earlier opinion, personality measures are related to job performance and do have a place in personnel selection.

A human characteristic which is difficult to define and which may be regarded as an aspect of personality is empathy. Empathy can be described as the capacity to put oneself in the place of another. This may take the form of vicarious experience of another's emotions or it may simply be a capacity for role-taking. Choplan *et al.* (1985) have concluded that the most appropriate way of considering empathy is to regard it as being multifaceted and containing both role-taking and vicarious experience components. In the agricultural literature, the term empathy has been used to describe the bond that exists between humans and animals under their care (English *et al.*, 1992). In fact an empathic bond may exist between a stockperson and his or her animals; however, empathy does not refer to the bond itself, which may have its origins in a number of factors, of which empathy is one. Empathy refers to that particular case where the stockperson may feel a bond with his or her animals because of being able to put him- or herself in the animal's position. Whether empathy is an important characteristic of a good stockperson is not clear. Some preliminary data from intensive farming industries will be reviewed later in this book.

1.5.3. Work motivation

The extent to which a person applies him- or herself to a task will depend, in part, on the extent to which the person 'wishes' to achieve the task. In other words, a good stockperson is one who is motivated to apply skills and knowledge to the management of the animals under his or her care. What this means, of course, is that the person must be motivated. In general, it is accepted that motivation alone is insufficient for good work performance; if a person does not have the knowledge, skills or opportunity to perform a job, motivation will not make any difference. However, if a person does have the knowledge, skills and opportunity to perform the task, what is the role of motivation in professional stockperson behaviour?

Motivation refers to the underlying forces that direct behaviour. Motivation cannot be directly observed, it is inferred from observed behaviour and, in Chapter 4, we discuss animal motivation in some detail in terms of the underlying state of the organism – particularly in terms of hunger or some hormonal state that directs behaviour. In the case of humans, there is considerable argument about the nature of motivation. On the one hand, motivation can be seen to result from the rewards and punishments that a particular behaviour has produced. In other words, motives develop through learning and depend on the history of rewards and punishments a person has experienced. How behaviours develop is discussed in some detail in

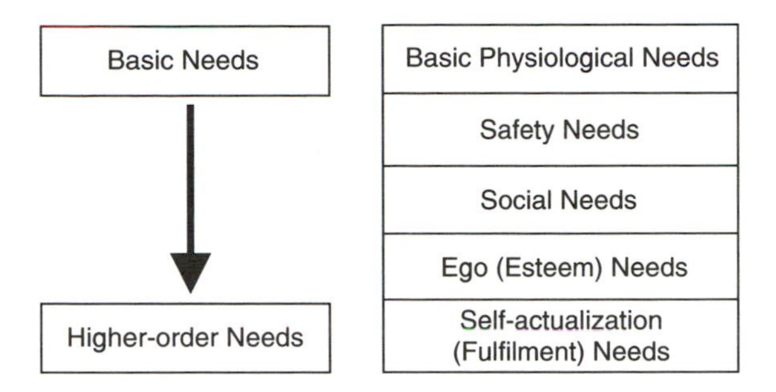

Fig. 1.2. Maslow's (1970) hierarchy of needs.

Chapter 5. This approach to motivation has its origins in the work by B.F. Skinner (1969). In the context of a job, rewards can include learned rewards, such as feelings of pride or accomplishment. At the other extreme, Maslow, a clinical psychologist, proposed that there is a hierarchy of innate needs which directs behaviour (Maslow, 1970). These needs, in sequence, were: physiological needs (air, food, etc.), safety needs, social needs (presence of others, etc.), esteem needs (recognition by others) and self-actualization (self-fulfilment). The idea is that people will be motivated by, say, a need for food or water but, if that need is met, they will be motivated by a need for security. If both of these are met, the person will be motivated by a need for social contact, and so on. Maslow's hierarchy of needs was based on a number of principles – in particular, the idea that the most basic needs were most fundamental to survival of the organism and that higher-order needs could be postponed without detriment to the organism (Fig. 1.2).

In some respects, it is not so important for us to settle on a particular theory of motivation as to consider the relevance of motivation, whatever its origin, to stockperson productivity. In fact, there is little systematic study of the effect of motivation on stockperson performance. However, a study carried out in India (Singh, 1983) showed that productivity, as measured by progressive farm behaviour, was associated with career interest, upward striving, attitude toward making money on the farm, intelligence, tolerance for work pressure and punctuality. This suggests that motivational factors, such as upward striving and need to make money, can contribute to productivity on a farm.

1.5.4. Absenteeism, job turnover and job commitment

Absenteeism is usually defined in terms of days absent and frequency of absences. However, it is desirable to distinguish between excused and unexcused absences. Absenteeism is one of the major sources of disruption and cost to industry, and agricultural industries are no exception. Anecdotal evidence from the pig industry in Australia indicates that turnover rates of 50% per annum are not uncommon. In the USA, Johns (1987) estimated that absenteeism costs to US industry at that time could be as high as $US30 billion with about 2–4% of the labour force being away from work on any given day. Despite the fact that most people mention some medical problem as the reason for absenteeism, it is unlikely that this is the real reason. There are a large number of factors that contribute to absenteeism, but the most consistent factor is level of job dissatisfaction.

Although the relationship between job dissatisfaction and absenteeism is consistent, it is not a particularly strong relationship. Jewell and Siegall (1990) have reviewed factors associated with absenteeism. The only personal factor that is consistently related to absenteeism is gender. Women are absent from work more than men. The explanation for this comes in part from the fact that women have family responsibilities that may require them to be absent. Another contributing factor may be that women generally occupy lower-status jobs than men, with lower levels of job satisfaction. Organizational variables, such as how boring the job is, the size of the organization and perhaps other characteristics of the work situation, may all contribute to absenteeism.

Turnover usually refers to workers leaving the organization and being replaced by others. However, it is sometimes the case that the organization may encourage the departure of a worker because of organizational change or because the performance of the worker is not satisfactory in some respect. Campion (1991) has defined turnover as individual motivated-choice behaviour. He argues that there may be a number of motivations underlying employee departure and, in order to satisfactorily analyse such data, it is essential that detailed records of employees' reasons for departure are obtained. Because good records of workers' reasons for departure are often not kept, it can be difficult to decide whether the departure was voluntary or involuntary. Job turnover is regarded as independent of absenteeism. In general, correlations between absenteeism and turnover are close to zero. Rusbult *et al.* (1988) suggest that decline in job satisfaction is associated with staff turnover. They argue that people with high levels of job satisfaction before it declined would be more likely to stay in the job and try to change the situation. Those who entered the job with low levels of satisfaction would be more likely to leave if satisfaction declined.

Most research has investigated the relationship between job satisfaction and turnover, and other variables have not received much attention. However, there is some evidence to suggest that it is the better performers who stay in the organization. Jewel and Siegall (1990) have reviewed the research in this area. They concluded that there were two key components which contributed to the intention of a worker to seek a new job. The first of these is the extent to which a worker is able to move to a new job. This capacity for movement is determined by the worker's task-relevant abilities, self-esteem and recent job-seeking experience. The second component is the worker's desire for movement to a new job. This is determined in part by the nature of the current job, the opportunity for progression in the orga-

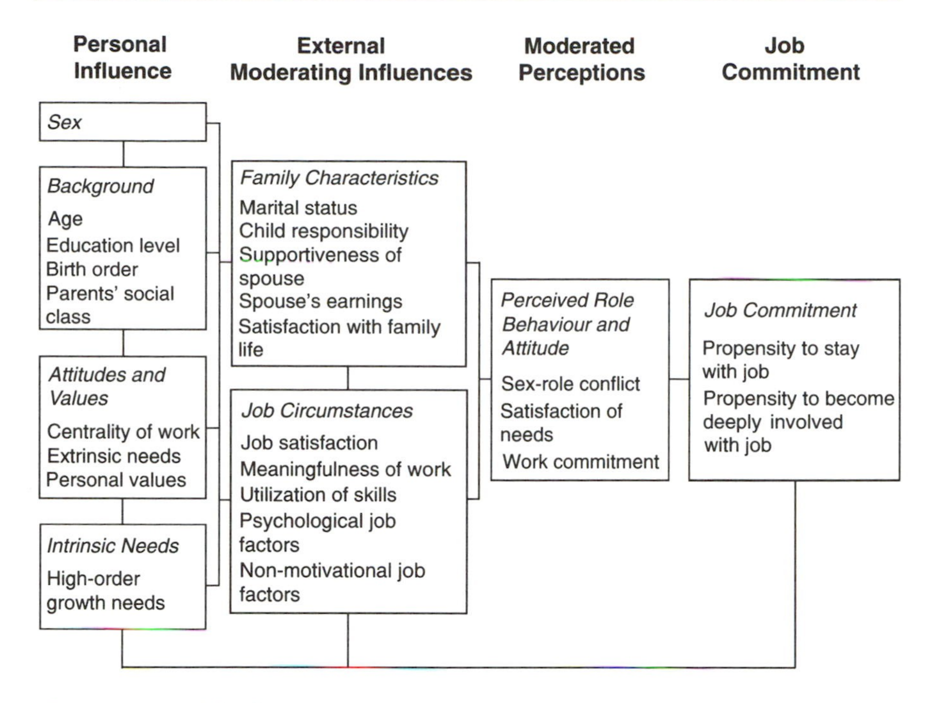

Fig. 1.3. A model of job commitment (adapted from Chusmir, 1982, with permission from the publishers).

nization and the rewards associated with the job. There appear to be no data on the causes of job turnover that specifically relates to agricultural industries.

Job commitment refers to the behaviour opposite to that of leaving the organization. Job commitment is a product of personal variables and the characteristics of the work situation.

There is evidence to suggest that people with a strong work ethic – that is, those who have a strong degree of job commitment – are also those who have a high degree of job satisfaction. Chusmir (1982) has proposed a model of job commitment in which personal factors, such as gender, age, educational level, work attitudes and needs, all contribute to job commitment (Fig. 1.3). These personal variables are moderated by family characteristics, including marital status, spouse support and earnings and satisfaction with family life. They are also moderated by job circumstances, including job satisfaction, meaningfulness of work, utilization of skills, psychological factors and other non-motivational job factors. All of these contribute, in turn, to the person's attitude towards the job, satisfaction of needs and work ethic, which together contribute to job commitment.

It is clear, therefore, that there is a complex set of factors that contribute to job commitment. There has been no systematic investigation of factors associated with job commitment in agricultural industries.

1.5.5. Job satisfaction

Job satisfaction is an attitude towards work based on how the person evaluates the work. In other words, job satisfaction refers to the extent to which a person reacts favourably or unfavourably to his or her work. In general, job satisfaction is thought to derive from the extent to which a person's needs or expectations are being met by the job. If job satisfaction is related to work performance, this provides a clear role for motivation in contributing, indirectly, to work performance.

Unfortunately, despite the widespread use of the construct in industrial/organizational psychology, well-substantiated theories of job satisfaction are difficult to find. Nevertheless, research shows that absenteeism and job turnover are associated with poor job satisfaction. In a review of studies from the Soviet Union, Phillips and Benson (1983) reported that there were numerous studies, both from the Soviet Union and from Western countries, showing a clear correlation between work dissatisfaction and the probability of switching jobs. They reported, in some detail, research by Ivanov and Patrushev (1976) into the 'social factors of increasing the productivity of labor in agriculture'. In general, 74% of farm workers reported that they were satisfied with their work, and job satisfaction varied directly with working conditions and skill level. This research showed that in the Moscow region about 30% of workers had decided or were almost resolved to switch jobs; 43.8% of young people in general were similarly disposed, as were 61% of the young people in animal husbandry. This research also showed a strong relationship between degree of work satisfaction and the desire to achieve better results. A strong inverse relationship was found between number of days missed and work attitude. A satisfied worker missed an average of 4.3 days each year, while a dissatisfied worker missed an average of 10.3 days annually.

1.6. EVALUATING JOB PERFORMANCE

Without doubt, the key feature of any employee is how well he or she does the job. There are two aspects to work performance – quality and quantity. Blumberg and Pringle (1982) have proposed a model of work performance that identifies three classes of contributing factors (Fig. 1.4). The first factor is capacity to do the job, the second is opportunity and the third is willingness. Capacity includes variables such as skills, health, ability and knowledge, while willingness includes motivation, job satisfaction and work attitude and opportunity includes working conditions, actions of coworkers and organizational policies and rules. This is not an exhaustive list of variables in Blumberg and Pringle's (1982) model, but it should be evident that all of the variables we have discussed so far contribute to work performance under the model.

As we have discussed in this chapter, there are some data in support of

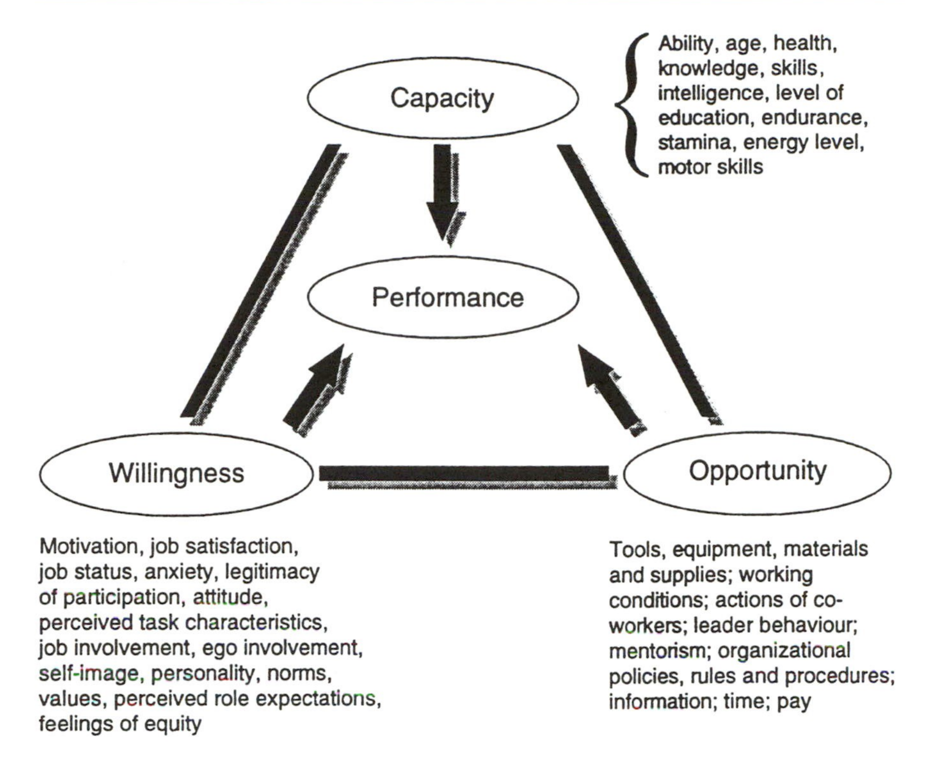

Fig. 1.4. A model of work performance (adapted from Blumberg and Pringle, 1982, with permission from the publishers).

the idea that these variables are related to work performance in agricultural industries, and there will be further review of relevant studies later in this book. Nevertheless, it should be clear by now that it is possible to characterize a stockperson in terms of the person's motivation, needs, work ethic, job satisfaction, abilities, skills and knowledge. Later in this book, a comprehensive model of factors contributing to the stockperson's performance with regard to animal productivity, animal welfare and farm productivity will be outlined.

1.7. THE STOCKPERSON'S ROLE: PREPARING THE STOCKPERSON FOR THE TASK

1.7.1. Training of stockpeople

If the question was asked of industry personnel about the role of the stockperson in agriculture, it would probably be generally stated that stockpeople have a critical role in livestock production. However, just how seriously and extensively this subject is taken and thus acted on is questionable.

Stockpeople often underestimate their value and contribution to agriculture (English *et al.*, 1992), and it appears that supervisors, managers, farm owners and industry leaders may also undervalue the contribution of stockpeople. Research and development in agriculture have focused on technological innovation, especially in areas such as housing, nutrition, breeding and health, and most of the industry training has generally targeted the training of supervisors and managers in these new technologies. In comparison, training of stockpeople has often been neglected, perhaps reflecting to some extent the attitude of senior industry personnel to stockperson training. The attitude of stockpeople to training and also the high staff turnover rates in some livestock industries may also be considered by some industry personnel to reduce the imperative for training.

The less-than-satisfactory opportunities for training stockpeople and the poor attitudes of many industry personnel to staff training also reflect a lack of appreciation of the importance of training in the industry. For example, while well over half of the employers and employees surveyed in a study of the Canadian pig industry recognized that specific skills were required to effectively conduct a number of routine tasks in pig production, surprisingly less than half of them generally believed that training was required to develop these skills (survey results quoted by English *et al.*, 1992).

Several authors have commented on the parlous state of stockperson training in a number of livestock industries. Lloyd (1974) reported that, while 86% of poultry stockpeople had no previous experience with poultry, only 4% of stockpeople had attended any form of training courses in the previous 2 years. Segundo (1989) found that 87% of 15 piggeries in Scotland had no off-site training available and 53% had no on-site training for stockpeople. Only 13% of Australian pig stockpeople surveyed by Kondos (1983) had received some organized technical training.

With the rapid technological development in agriculture, it is imperative that technological training of its staff should occur. This training, which should contain both the appropriate theoretical components and practical training, should target all farm staff, not just senior staff, and should be an ongoing process. For stockpeople this training obviously needs to focus on technical skills and knowledge, occupational health and safety and equipment use and maintenance, and perhaps, where appropriate, on interpersonal skills, such as working effectively in teams. The widespread availability and the effectiveness of both on-site and off-site training of stockpeople are highly variable.

Where available, agricultural training for stockpeople has been traditionally college-based, with the focus on classroom teaching and often less than ideal opportunities for experience in commercial-like conditions. One could also question the relevance of some of these programmes in addressing the requirements of industry and stockpeople themselves. Many of these courses were particularly aimed at future supervisors and managers, with little relevance to base-grade stockpeople. Apprenticeship schemes offer

Fig. 1.5. On-site training of new staff is becoming increasingly common in the livestock industries.

both theoretical and practical training, but at times the focus is on training future supervisors and managers. Many livestock industries in many countries are well serviced by industry days and seminars offered by government or university agricultural agencies and agricultural-product companies, but these programmes often target senior farm staff.

There appears to be a growing international trend towards providing more training for stockpeople on site (Fig. 1.5) and less in agricultural colleges or technical-education centres. The costs and logistics involved in replacing staff and releasing staff to travel to training venues may certainly influence industry's support of this development. On-site training may also be more attractive to industry because it occurs in a commercial setting, in which relevant practical sessions can be offered, and because the training can occur in a familiar, less formal, training setting, which is often in contrast to the more formal classroom setting. This may be particularly relevant since the latter setting may be more threatening and intimidating for many base-grade stockpeople, who may have previously had difficulty with a formal educational system.

Organized discussions between experienced and inexperienced staff have and continue to be a common on-site training activity (Cleary, 1990). However, large corporate farming companies appear to be becoming increasingly active in developing and introducing local outcome-orientated training for their stockpeople. The on-site training may vary from simple training activities in which the more experienced staff, not necessarily with any formal instruction on teaching techniques, act as mentors for inexperienced staff to the situation where structured in-house training programmes, conducted either by experienced staff or trained human-resource management staff, are undertaken and these training activities are linked with competency assessments and remuneration (Miller, 1995). Indeed, some of these in-house training schemes are very progressive, with well-developed syllabuses and involving both classroom instruction, by either local or external trainers, and skills training under commercial conditions.

Low levels of training skills by senior farm staff can inhibit learning in training programmes conducted on site, and a lack of suitable reference material for in-house training appears to be a limitation for many livestock industries. Staff cultural attitudes about education may also inhibit learning in both on-site and off-site settings. Multimedia training packages, in which information is presented in several forms and which allow stockpeople to individually interact at their own pace with the teaching programme in an informal non-threatening manner, may be more conducive to learning. This approach also has the potential to relieve managers or senior stockpeople of formal didactic teaching sessions and allow them to be used more effectively as facilitators in group sessions, where experiences can be shared and views developed in an advantageous setting. Furthermore, these interactive multimedia programmes allow staff to be released for training without major disruptions to work and to progressively complete training modules and be assessed at times that are convenient for management. Support by government and commercial services in the provision of interactive, user-friendly, multimedia packages on key technical issues and the provision of trainers or opportunities to train in-house trainers to facilitate the training and assess staff undergoing this training, will make training more accessible to industry.

1.7.2. Selection of stockpeople

Selection is the process of matching a person to a particular job. This matching process can involve assessing applicants with respect to many of the variables discussed earlier in this chapter. For example, because work preference, skills, knowledge, work ethic and, perhaps, personality factors may all contribute to worker performance, high job satisfaction and capacity for development in the job, it may be appropriate to measure all of these factors as part of a selection process. However, in general, demographic data and previous experience may also be useful in screening out potential employees with a low chance of success. Usually, a face-to-face interview with an applicant can also provide useful data if the interviewer is experienced in selecting people for a particular industry or profession. As an example of the advantages of a selection procedure, Borofsky and Smith (1993) found that the use of a selection procedure resulted in a reduction in turnover, accidents and absenteeism.

It is important that the person carrying out the selection process should behave ethically towards job applicants. There are three basic facets to this responsibility. First, the selection officer should ensure that the selection processes that he or she uses are as up to date and reliable as possible. Second, information provided by the applicant should be kept confidential and should not be used for any other purpose without explicit permission from the applicant. Finally, the selection officer should evaluate the efficacy of the selection procedure and ensure that the organization is able to accurately determine the costs and benefits of the selection procedure.

At present, there are no validated procedures for selecting stockpeople in agriculture and, as we shall see later, despite the fact that we can identify some characteristics that affect productivity in intensive-farming industries, this knowledge has not been incorporated into a selection procedure. Some suggestions to indicate how this might be done will be discussed later in the book.

1.7.3. Human-resource issues

In non-agricultural areas, the need to retain desirable staff and to foster high standards of work ethic are generally well recognized and addressed. The management of the human resources in agriculture at the level of the stock-person appears to be highly variable. The earlier comments about lack of recognition of the stockperson's role in animal performance and welfare, the small-scale operations that are often prominent in agriculture and the lack of business management skills seen at times in agriculture may contribute to this weakness.

One of the minor aims of this book is to foster recognition of the role that the stockperson plays in agriculture. Recognition of this contribution should facilitate improvements in stockperson selection and training and appropriate financial and personal rewards for stockpeople, all of which should contribute to higher job performance and staff retention rates. The opportunities for staff selection and training are explored in detail in the latter half of this book, after the interrelationships between the characteristics of stockpeople and the behaviour, performance and welfare of farm animals have been reviewed.

1.7.4. The stockperson and animal welfare

Because of the important influence of the stockperson on the welfare of farm animals and the widespread community interest in animal welfare, its measurement and its consequences, Chapter 2 discusses some of the welfare issues that are critical to the sustainability of livestock production. Indeed, general community concern about animal-welfare problems in a livestock industry, through consumer boycotts or government regulation, may influence the ability of the industry to produce or sell its animal products. It is important to discuss these issues in a book about stockpeople, because the stockperson plays an integral role in safeguarding the welfare of both intensively and extensively farmed animals. The importance of this topic should not be underestimated because of its implications not only for the animal but also for the sustainability and economics of the livestock industries.

Chapter 2

The Ethics of Animal Farming: Implications for the Stockperson

2.1. INTRODUCTION

There is considerable public interest in animal welfare and most people believe that animals, including farm animals, should be not be subjected to pain or severe discomfort (Fraser and Broom, 1990). The livestock industries are also sensitive to the issue of the welfare of farm animals. A deterioration in the welfare of animals is often associated with reductions in individual animal performance. Furthermore, local and international agricultural markets are fiercely competitive and, in addition to the necessity of technological improvements, development of new products and marketing expertise to maintain competitiveness and increase sales, livestock industries need to project a welfare-friendly image of their products to maximize their marketing advantage. Concerns about the welfare of farm animals in a particular industry may influence the buying behaviour of potential consumers of the product from that industry. Codes of practice or government regulations may restrict specific practices in a particular livestock industry that the general community finds objectionable on welfare grounds. The image of a welfare-friendly product requires farming practices that minimize the risk to animal welfare and the provision of objective information that positively influences the consumers' beliefs about the welfare implications of the farming practices that produce this product. Indeed, the results of welfare research on farming practices will influence both industry practices and the consumers' perception of the product.

While not widely recognized, stockpeople have a critical role in ensuring that the welfare of their livestock is not compromised. The general public and indeed many within agriculture would contend that housing systems pose the main risk to animal welfare in intensive-farming systems; however, the contribution of stockpeople to animal welfare in both intensive and extensive systems is poorly recognized. This chapter introduces the topic of the human–animal relationship in agriculture and its consequences for the welfare of farm animals. Because of its implication for the animal, the stockperson, the livestock industries and the general public, a detailed discussion of animal welfare is presented in this chapter. This discussion of animal welfare focuses on its definition, its assessment and its consequences for the animal.

2.2. HUMAN–ANIMAL RELATIONSHIPS IN AGRICULTURE

Humans working closely with farm animals develop relationships with their animals often not dissimilar from those that develop between humans and companion animals. The similarity of the two relationships, particularly in terms of their strength and quality, may surprise many people. The fact that most of us view our pets as companions, often with considerable affection, is well recognized. However, while the symbiosis or interdependence of this relationship between humans and animals in agriculture is recognized to some extent, the general public probably does not recognize the general fondness and friendship that many farmers or stockpeople have for their animals. The general public's romantic view of livestock farming involving the shepherd looking after the 'flock' does not extend to modern farming systems, and in fact the human–animal relationship in modern farming is often viewed by the general public as an exploitative one by humans, in which little or no regard is afforded to the welfare of these farm animals. This is a simplistic and inaccurate way of describing the human–animal relationship in modern agriculture, because it fails to recognize the interdependence between farmers and their animals.

As in the early phases of domestication, the relationship between humans and animals in modern agriculture has components of symbiosis, in that, in return for the animal products that they provide for humans, the animals are maintained and cared for by humans. Undoubtedly, the satisfaction of economic demands is the main feature of this human–farm animal relationship today. A similar, needs-based, relationship often exists between humans and pets in the general community. For example, caged birds and fish in aquaria in many households are predominantly kept for entertainment and aesthetics. Despite the fact that reciprocal need has been, and still is, a feature of the relationship between humans and domestic animals, our ethics concerning the care and use of animals in general has greatly improved in recent times.

The popular view of modern farm animals being exploited has probably arisen for several reasons. Economic pressures have led to intensification of farming, with large numbers of animals under tightly controlled environmental conditions and managed by few stockpeople, who often may be employed with limited animal husbandry experience or training. This intensification of farming was designed to optimize animal growth, reproduction and health. This trend towards intensification has probably been influential in shaping public opinion. Unfortunately, at times, these management innovations, particularly the housing developments, were introduced without adequate knowledge of or regard for the animal's ability to adapt to these changes, and consequently animal welfare has sometimes been compromised. Furthermore, the general public's requirement for cheaper farming products has dictated the search by agricultural scientists for maximum animal output at minimum cost, which invariably has promoted a view of exploitation of animals. The enormous difference today in the lifestyles and experiences of people in urban communities compared with those in rural communities does not assist the general public's awareness and understanding of what has happened and is happening in agriculture, which, in turn, often creates suspicions about conditions for farm animals. The farming communities around the world can certainly improve their efforts in educating the public about modern farming practices and the reasons for these practices.

Most farmers recognize that deterioration in the welfare of their animals may result in depressions in the productivity and health of individual animals, with potential adverse consequences for profitability. Farmers obviously consider their animals as resources, but farmers have long treated and viewed their animals with affection as companions (Fraser and Broom, 1990). This opinion may be difficult for some outside agriculture to accept, and we shall return to this point later in the book. The quality of these human–animal relationships, measured in terms of the farmer's beliefs about their animals and their behaviour towards these animals, as well as the behaviour of these farm animals in the presence of humans, is considered in detail in a number of animal industries.

It is indisputable that there have been cases of animal abuse and cruelty in agriculture, just as there have been cases involving companion animals, and these are likely to continue. However, recent developments, particularly in terms of awareness and recognition of the subject, continuing research and the introduction of improved codes of practice, indicate the promise of ever-increasing improvements in farm-animal welfare. A similar situation exists with the welfare of companion animals. Improvements in a range of areas, such as education of the community in terms of its responsibility, suitability of particular species as pets and the proper care and maintenance of pets, are likely to reduce the incidence and magnitude of abuses that have occurred in the past. Community pressure on the welfare of laboratory animals has also resulted and is likely to continue to result in improvements in the care and use of these animals.

While physical, social, nutritional, disease and climatic factors may influence the welfare of farm animals, the competency and motivation of the stockperson in the care and management of the animal are critical ingredients in determining the welfare of farm animals. Ultimately it is the stockperson that is charged with ensuring that a particular farming system is operated properly and diligently. Weaknesses in the motivation of the stock-person to follow farm protocols for animal care and maintenance and to monitor and promptly address production and welfare issues arising in his/her area of responsibility will place animal welfare and productivity at risk. Furthermore, the large number of animals under the care of an individual stockperson in modern farming systems, together with any weaknesses in the competency and diligence of an individual, may have serious effects. This arises, not only because of the number of animals involved, but also because of the almost total dependence of animals on humans in some of these systems. We shall contend throughout this book that the attitude of the stockperson towards his/her animals, a factor for which the effect on animal welfare has in general been ignored, is highly influential in determining animal welfare and productivity in modern agriculture. As the dependence of animals on the stockperson for their care and maintenance increases, the influence of the attitude of the stockperson on the welfare and productivity of modern livestock correspondingly increases.

2.3. ANIMAL WELFARE AND THE DEBATE

Animal welfare is a highly emotive subject and most of us are not spared the emotions that the topic can create. Irrespective of our lifestyle, most people in modern society are regularly confronted with cases of animal suffering. Images in wildlife documentaries portraying the survival of the fittest individuals and the death of the weakest generally strike an emotive chord in most of us. The news media are not reticent in extensively covering animal-welfare abuses because of the public interest that these stories generate. Most people experience firsthand, at some stage in their life, the suffering of an ageing or ill pet and the difficulty in assessing the pain or suffering of a loved animal. The emotional effects of our possible loss may accentuate our concern. In general, we feel a degree of empathy for our pets; we tend to think of their suffering in human terms and try to deal with the situation much as we would with another human. In fact, there is abundant evidence in the literature for pets being treated as a family member (Voith, 1985; Cain, 1991), even to the extent of jealousies and conflict arising because the pet competes for our affections (Cox, 1993).

Nevertheless, animal welfare is a controversial subject because of its subjectivity. A range of views on the subject, often based predominantly on value judgements, exist within the general community, leading to marked and often extreme attitudes on animal-welfare issues. The action of some welfare groups in lobbying and boycotting specific animal industries or practices is indicative of their views and the strength of these views. Equally, there are strong community views about the adverse effects of domestic animals. Dog bites receive wide publicity and, in Australia, so does the predation on native wildlife by domestic cats. This, in turn, leads to legislative change restricting the movements of these animals, with possible welfare implications. Decisions about animal welfare are obviously morally and politically important. Failure to address the community's views may result in the general community, via governments, restricting people's use of and access to animals.

The public interest in the welfare of farm, laboratory and companion animals has rapidly increased in Western society during the last 30 years. The writings of a number of authors (e.g. Harrison, 1964; Singer, 1975) were important in raising this as a topic in the general domain, and the ensuing interest and debate clearly focused, for example, the farming community and, in turn, the scientific community on the topic of animal welfare. Most people, irrespective of their involvement in agriculture, would probably agree that the debate has had positive effects on the welfare of farm animals and, indeed, animal welfare is a topical subject for those with interests in livestock production.

2.4. THE COMMUNITY'S VIEWS ON ANIMAL-WELFARE ISSUES

Most people accept that humans have a moral obligation towards farm, companion and laboratory animals. In addition to the undeniable benefits that these animals provide for humans, the domestication of these animals has increased their dependence on humans and thus necessitates this obligation. However, what is at question for most people is the extent of this obligation, particularly in relation to the standards of welfare that society should provide for these animals. A consensus on this aspect of fair and humane treatment is difficult to achieve when such diverse views exist in the general community. In some countries, specific welfare codes for farm, laboratory and companion animals are prescribed in legislation, while in most Western countries codes of practice have been developed that can be used in conjunction with legislation relating to cruelty to animals to deter and prosecute cases of welfare abuse. However, in many countries such protection generally does not exist.

In making a decision on whether or not an animal's welfare is seriously compromised, individuals will integrate moral views with biological facts. Biologists thus have the important role of establishing the biological facts on how animals biologically respond to the practices in question, whether they are farming, laboratory or general community practices regarding animals. While assembling and gaining a consensus among scientists on the biological facts associated with a specific welfare issue would appear to be easier to achieve than gaining a general view within the general community on a welfare issue, this is not necessarily true. As will be discussed shortly, conflicts have arisen in scientific circles for several reasons: definition of welfare varies among scientists, welfare research requires a multidisciplinary approach, only limited research has been conducted to date, and the results of this welfare research are often complicated and, at times, appear contradictory. This is a serious limitation, since one of the important steps in developing defensible policies on animal care and use is to assemble factual information on the animal's biological changes under the particular system or treatment. Without factual information, it is difficult for a society to develop a consensus on a defensible policy on an animal-welfare issue.

Although science has an important role in providing sound defensible information on how animals respond to a specific practice, ultimately it is an ethical decision by the general community that will determine the acceptable welfare standards for farm, companion and laboratory animals. However, the development of a clear consensus within a country on an ethically and scientifically defensible philosophy on animal welfare is obviously difficult. A society's attitudes to the use of and obligations to farm, companion and laboratory animals are extremely disparate, influenced by demographic factors, religion and culture and varying over time with economic and ideological changes. As Teutsch (1987) has reported, there are clear within- and between-country variations in the attitudes of people to their obligations towards animals.

The debate in the general community about the care and use of animals in agriculture, research and modern society has increased over time and there is continuing widespread pressure, particularly in developed countries, for legislation and regulation that will protect animals from pain and distress. The problem in developing such guidelines and regulations is to define what constitutes good welfare or well-being for the animal. Without such a definition, legislation concerning the care and use of farm, companion and laboratory animals will be developed based on emotions, without serious regard for objective data, and, notwithstanding the good intentions of most people involved in the process, could jeopardize animal welfare.

2.5. DEFINING AND MEASURING ANIMAL WELFARE

Because of the widespread use of the term welfare in a number of scientific disciplines, in philosophy and in the general community, definitions of welfare vary considerably. The term is often used to avoid being too specific about the nature of the particular issue (Broom and Johnson, 1993). Without a clear definition, welfare cannot be studied, because it cannot be measured either directly or indirectly. Furthermore, a definition is required to provide clarity and precision in use in legislation and codes of practice. Ambiguity of the definition creates considerable problems in developing a useful debate on this topic in the general community.

Scientists have attempted to add objectivity to the debate; however, for a number of reasons, research results are often complicated and, at times, both the results and the conclusions appear to be contradictory (Mason and Mendl, 1993). Four of the main reasons for this apparent contradiction are

Fig. 2.1. Implanted catheters enable blood samples to be collected quickly and simply without influencing the corticosteroid concentrations of the samples. The pigs in this photograph have indwelling catheters and the collars around the animals' necks protect and store the external section of the catheters and their attached taps.

differences between scientists in their definitions of animal welfare and in the methodology used to quantify animal welfare and a lack of fundamental knowledge of the variables studied and the proper measurement of the variables under study. The last point is particularly true in a number of studies that have attempted to utilize changes in stress physiology to assess welfare (Barnett and Hutson, 1987). It is useful at this stage to use some of the problems associated with measuring physiological changes as examples of these methodological problems. Both the surgery to implant indwelling catheters (Fig. 2.1) for subsequent blood collection and the blood-collection procedures themselves may influence the physiological variables being studied. For example, surgery and, in particular, general anaesthesia will affect the hypothalamus–pituitary–adrenal axis (which is the physiological substrate that mediates the stress response) and thus failure to allow experimental animals to fully recover from surgery may confound the effects of treatment on physiological measures of stress, such as basal corticosteroid concentrations. Brief handling and restraint of animals to allow a venepuncture blood sample to be collected may induce an acute stress response and therefore may mask basal corticosteroid concentrations in experimental animals. The episodic nature and the diurnal pattern of corticosteroid secretion require regular sampling, with minimal disturbance to the animals, to provide reliable data on basal concentrations. Thus, inappropriate and erroneous sampling procedures to monitor acute and basal values of the physiological variables and confusion about the value of acute versus chronic stress responses to assess welfare may give rise to misleading conclusions.

Most recognize that one problem in assessing welfare is the inherent inadequacy of the abstract concept of welfare (Barnett and Hutson, 1987). The problem is difficult to solve when definitions of welfare include 'well-being', 'suffering' or other subjective 'feelings'. However, the lack of an objective definition of welfare does not lessen the importance of the concept.

There is general acceptance within scientific circles today that assessment of animal welfare relies on a multidisciplinary approach. During the last decade, there have been an increasing number of studies that have utilized changes in more than one of the following as evidence of risk to animal welfare: changes in stress physiology, behaviour, immunology, injury, morbidity, mortality and productivity. Productivity of the individual animal, such as growth rate or reproductive performance, is often included, since a stress response, particularly a chronic stress response, is likely to depress individual animal productivity. Disease, injury, movement difficulties and growth abnormalities are indisputably indicators of poor welfare; however, data on the incidence of these conditions that may indicate acceptability or unacceptability are not available (Broom and Johnson, 1993).

Apart from evidence on morbidity or mortality, changes in biological responses will not provide an absolute assessment of welfare. Welfare research with our current knowledge relies on a relative assessment, in which the biological responses of animals to a particular treatment are compared with those of a control. That is, does the treatment under study lead to better or worse biological responses than those of the control and thus does the treatment pose a lesser or greater risk to the welfare of the animal?

2.5.1. Welfare assessment using biological responses and their consequences for the animal

An approach that has been advocated by a number of authors (Moberg, 1985; Barnett and Hutson, 1987; Barnett and Hemsworth, 1990) is to measure the stress response of an animal to assess the animal's welfare. As Moberg (1985) contends,

> it is not difficult to argue that if an animal is stressed, its well-being is threatened. On the other hand, if the animal is living under conditions that it finds non-stressful, it is questionable that the animal's well-being is truly threatened.

This approach relies on measuring the biological responses of the animal to an environmental change (external stimulus) that may challenge its homoeostasis. Homoeostasis refers to maintenance of the internal state within certain limits, which is necessary for the animal to survive. Thus introduction of a new housing system or husbandry procedure may challenge homoeostasis, and the biological responses that the animal makes in its attempts to maintain homoeostasis constitute the animal's stress response. The main biological responses available to the animal to cope with a stressor are behavioural and physiological responses and these will be considered in more detail later. Others have supported a similar approach. For example, Broom (1986) defined the welfare of an individual as 'its state as regards its attempts to cope with its environment'.

Central to this approach are the behavioural and physiological stress responses by the animal to a putative stressor. Implicit in this approach is the search for a biological cost to the animal as a consequence of either behavioural or physiological responses to the environmental change. Thus investigations using this approach seek evidence of behavioural and/or physiological change that may have serious biological costs for the animals. For example, exposure to a stressor that produces behavioural changes, such as cannibalism or self-mutilation, which in turn adversely affects the morbidity or mortality of the animal clearly has serious consequences for the animal. Vices, such as tail- and ear-biting and belly-nosing in pigs are generally viewed as abnormal behaviours, indicative of reduced welfare due to the risk of injury or at times death. Similarly physiological changes, such as a prolonged activation of the hypothalamus–pituitary–adrenal axis, leading to a suppression of the immune system or depressions in growth or reproduction, have serious consequences for the animal. The magnitude of this biological cost is a reflection of the seriousness of the environmental challenge to the homoeostasis of the animal.

A relatively recent but subtle development of the above approach is the view that welfare can be considered within the concept of biological fitness (Fraser and Broom, 1990; Broom and Johnson, 1993; Hemsworth *et al.*, 1996b). This concept of biological fitness generally applies to natural populations and refers to 'fitter' animals having a greater genetic contribution to subsequent generations (Pianka, 1974); this is based on their abilities to successfully survive, grow and reproduce. While the last attribute may not always apply to individual farm animals, since reproduction is either controlled or absent for many farm animals, the ability to grow and survive could be considered measurements of 'fitness' within the limits of the management system. Most production systems in agriculture have breeding and growing components and these can generate considerable data on the reproductive success of individuals. For example, conception rates and mortality, morbidity and growth of offspring can be used as a measure of 'fitness'. Similarly, Beilharz and Zeeb (1981) and Beilharz (1982) have linked reproductive performance of domestic species with welfare.

This approach of using the biological fitness of the individual to assess welfare recognizes that there is a range of behavioural and physiological responses available to the animal in its attempts to cope with an environmental change that challenges homoeostasis. It is these behavioural and physiological responses that may have serious consequences for the fitness of the individual. As with the previous approach, this approach utilizes not only the magnitude of the biological responses to the stressor, but also the biological costs to the animal of resorting to these responses, to assess welfare. Clearly, the most serious risks to animal welfare are those situations involving long-term exposure or frequent exposure to a stressor that results in the animal experiencing a chronic physiological stress response. For example, housing in overcrowded conditions or housing on tethers results in a prolonged activation of the hypothalamus–pituitary–adrenal axis in pigs, with immunosuppression and depressions in reproduction (Barnett *et al.*, 1985, 1987; Hemsworth *et al.*, 1986b; Barnett and Hemsworth, 1991). Behavioural change as a consequence of a chronic stressor may also affect fitness. Dietary deficiencies in farm animals can lead to chewing of wood, soil, hair or faeces, with the possibility of adverse consequences on fitness through the ingestion of some of this material (Fraser and Broom, 1990). Similarly, feather-pecking of group-mates will influence the fitness of poultry, as will ear- and tail-biting of group-mates affect that of pigs. The ability to demonstrate both significant biological responses, such as prolonged behavioural and physiological changes, to a stressor and consequent depressions in fitness obviously increases our confidence in interpreting a serious risk to the welfare of the animal when exposed to this particular stressor.

While it is generally recognized that the most serious risks to welfare are those stressors that are imposed on the animal in the long term, either continuously or frequently, the magnitude of acute responses in practice can be used to assess the intensity of the stressor (Barnett and Hutson, 1987). This view is predicated on the idea that the magnitude and duration of an acute stress response reflect the intensity of the stressor and thus can be useful in differentiating between the relative aversive or noxious components of stressors under comparison. For example, studies on the effects of transport indicate that sharp increases in the corticosteroid concentrations of calves occur immediately after loading and unloading. The concentrations fall gradually as the journey progresses, indicating that the loading and unloading of the calves are more stressful than the actual journey (Kent and Ewbank, 1983, 1986a, b; Fell and Shutt, 1986). On the basis of respiration rate and corticosteroid responses to 10-min and 2-h transport, Tennessen *et al.* (1984) concluded that cattle found the handling and loading aspects of transport to be more stressful than the transport itself. While increases in heart rate, respiration rate and cortisol concentrations would be expected in response to transport, they can be considered as part of the 'normal adaptive response' of the animal to a novel situation. In themselves, these physiological changes are difficult to interpret in terms of welfare. The main value of these physiological criteria lies in their ability to identify the manipulations of the procedures involved in the transport process that cause minimal changes in these variables for the shortest possible time. Thus, they can be used to develop procedures that minimize potential risks to welfare; for example, they can be used to determine the effects of ramp angles for loading and unloading transported animals. Furthermore, there is limited evidence that acute stress responses of sufficient magnitude for sufficient duration may affect reproduction, growth and immunology (Kelley, 1985; Klasing, 1985; Moberg, 1985), and thus may have implications for fitness.

The approach of considering welfare under the concept of biological fitness is a subtle extension of the approach of using a stress response and its consequences for the animal to assess welfare. The former approach has introduced a definition of welfare in terms of biological fitness, which brings with the concept indices that are well recognized for natural populations but can be extended to domesticated populations. There are some difficulties with this approach, such as humans controlling the growth and reproductive stages of farm animals (for more details, see Broom and Johnson, 1993). However, efficiency of growth and reproductive performance within a standard production system where the inputs are controlled can be used as fitness variables, as components of the system that is being assessed are manipulated in a controlled scientific manner. The concept of stress has been the subject of some controversy within scientific circles, but particularly for those scientists interested in animal-welfare assessment, and the concept of stress and its measurement will be considered later. Nevertheless, the approach of examining the animal's biological response to assess welfare clearly reflects our current state of knowledge and will be utilized in this book.

In order to utilize this approach, it is necessary to understand the behavioural and physiological responses that animals may use in addressing an environmental change and how these changes may be studied. The next two sections deal separately with these biological responses, but it should be appreciated that, through the central nervous system, these responses are integrated.

2.5.2. The concept of stress

The contribution of Selye (1946) to the development of a comprehensive theory of stress has been substantial. Although some key elements of Selye's (1946) general adaptation syndrome (GAS) have been reviewed and amended, his contribution has been critical to the development of our current understanding of the animal's response to stressors. It is now clear that stress responses are not restricted to the hypothalamus–pituitary–adrenal axis and, furthermore, the role of other endocrine pathways and the implication of the psychological component of the stressors with regard to the biological responses of the animal are now recognized.

As described by Selye (1946), the GAS involves a series of non-specific biological responses by the animal to a range of diverse stressors, and the effects of these non-specific responses are viewed as cumulative. Thus, Selye considered that these biological responses were not dependent on the type of stressor, whether they were physical, climatic or social, and basically three stages of response were described: an alarm reaction, resistance and exhaustion. The initial stage included the activation of the hypothalamus–pituitary–adrenal axis, the second involved adaptation, in which the animal is physiologically capable of coping with the stressor, and, if the challenge remained, the animal would reach an exhaustion stage, in which its ability to resist/adapt declines. In this latter stage, in a matter of months there would be exhaustion of the biological defence system of the animal. Selye argued that during this final stage, pathologies would develop as a consequence of prolonged activation of the adrenal–pituitary axis.

The results of research by Mason (1968a, b, 1971) challenged Selye's (1946) view of the non-specific response of the hypothalamus–pituitary–adrenal axis to stressors. This research indicated that, when the psychologically threatening or arousing aspects of the situation were reduced, stressors such as fasting or elevated temperatures had less activating effects on the hypothalamus–pituitary–adrenal axis. For instance, gradual increases in temperature resulted in reduced concentrations of corticosteroids, while fasting with the provision of non-nutritive flavoured pellets did not elevate corticosteroid concentrations. Elevated temperature is a situation in which activation of the hypothalamus–pituitary–adrenal axis would not be biologically adaptive, because such a response would use energy and increase body heat. A number of studies have also shown that exposure to novelty or uncertainty is one of the most potent activators of the hypothalamus–pituitary–adrenal axis (Levine, 1985). Experience with an aversive stimulus (stressor), particularly in terms of the animal's ability to predict and control the stimulus, will also markedly influence the animal's response to the stimulus (Weiss, 1971). As recognized by a number of authors, the mode of response and the magnitude of the response are determined more by how the animal perceives the stressor rather than by the physical characteristics of the stressor (Moberg, 1985). These views are not inconsistent with Selye's (1946) ideas.

Although there is considerable argument about the value of Selye's GAS and perhaps even its contribution to our understanding of stress, there are significant elements of many of our current views on stress that have been retained or derived from Selye's early views on the biological response of animals to stressors. The following is a description of stress that will be utilized in this book, and this description, which relies heavily on Selye's views but with refinements that have arisen from subsequent research, is similar to that used by a number of authors (Moberg, 1985; Barnett and Hutson, 1987; Broom and Johnson, 1993).

There are basically three types of interrelated biological responses that are available to an animal when confronted with an environmental change (stressor): behavioural responses, responses of the autonomic nervous system and responses of the neuroendocrine system. The central nervous system integrates these responses to provide the animal with the principal resources to cope with the stressor. An example that is commonly quoted to demonstrate how the central nervous system integrates these responses is the following, involving a rat that is placed in a cold environment. The rat will attempt to maintain its body temperature in several ways. It may respond behaviourally by seeking shelter or building a nest and physiologically by reducing peripheral blood flow and increasing metabolism.

One of the earliest recognitions of the animal's biological response to an environmental change was the description by Cannon (1914) of the autonomic nervous system's response to a stressor. This response which he labelled the 'flight-or-fight' syndrome, is characterized by a rapid, specific response by the autonomic nervous system and secretions of catecholamines (adrenaline (epinephrine) and noradrenaline (norepinephrine)) from the adrenal medulla. These responses function to mobilize the body's reserves to cope with the challenge, such as an increase in heart rate and blood flow and a set of metabolic changes, particularly the production of glucose from liver glycogen.

These 'fight-or-flight responses' last for only a short period and, if the stressor or challenge is not removed, a second series of events occurs. This is the short-term or acute stress response and is corticosteroid-dependent (Selye, 1946, 1976). Corticotrophin-releasing hormone (CRF), released from the hypothalamus, stimulates adrenocorticotrophic hormone (ACTH) release from the pituitary, which, in turn, stimulates the release of corticosteroids or glucocorticoids from the adrenal cortex. Arginine vasopressin (AVP) from the hypothalamus has a role in some species in stimulating ACTH secretion. This acute response may last from minutes to hours and has the major function of providing glucose from non-carbohydrate sources (particularly protein from muscle) for the required increased metabolic performance. If the stressor is removed, this physiological state will disappear, with possibly no real ill effects on the animal apart from a depletion of energy reserves. However, it must also be recognized that, while acute stressors are short-acting, they could have detrimental effects. For example, while a single event of an acute stress response may not be detrimental, it is unknown what magnitude and duration an acute stress response or a series of acute stress responses would need to be before there were adverse effects. There are a number of examples where an acute stress response at specific times in the reproductive cycle has interfered with different aspects of reproduction (Moberg, 1985; Rivier and Rivest, 1991; Clarke *et al.*, 1992). Because of the importance of the series of carefully orchestrated endocrine events required for oestrus, ovulation and conception and the known effects of stress on these endocrine events, it is perhaps not surprising that activation of the hypothalamus–pituitary–adrenal axis prior to mating may adversely affect female reproduction.

If the stressor remains (i.e. a prolonged stressor), the response continues to the third series of events, which is the long-term or chronic stress response. This response is also corticosteroid-dependent and comes at a physiological cost to the animal (i.e. decreased metabolic efficiency, impaired immunity and reduced reproductive performance). Research on a number of species has shown that chronic stress can have detrimental consequences on growth and reproduction. Furthermore, chronic stress can have detrimental long-term influences on health, with such effects as ulcers, hypertension, arteriosclerosis and a suppression of the immune system (Moberg, 1985; Barnett and Hutson, 1987). How serious these costs are depends on how long the animal is required to divert physiological resources to maintain homoeostasis.

While the role and actions of corticosteroids in acute and chronic stress responses are well known, this is not to imply that corticosteroids are the only physiological variables affected by stressors. Other hormonal systems are responsive to stressors in a number of species, and changes have been identified in a number of hormones, including catecholamines, thyroid hormones, growth hormone, prolactin and endorphins (Selye, 1976; Levine, 1985; Broom and Johnson, 1993). However, our understanding of the significance of some of these changes is poor.

2.5.3. Behavioural change

Early attempts by scientists to assess welfare mainly used behavioural change, particularly the occurrence of unusual or abnormal behaviours, as primary evidence of reduced welfare, without consideration of other biological changes and the consequences for the animal.

Behavioural change is often the first response by an animal in its attempts to cope with an environmental change and, indeed, behavioural change may be an effective strategy in successfully coping with the stressor without having to resort to a physiological change. For example, avoidance or the display of submissive postures in the presence of a dominant animal is an appropriate response by a subordinate animal to a threat by a dominant animal. Similarly, wallowing and separation when lying are appropriate behavioural responses for pigs when temperatures are elevated, while huddling is appropriate when temperatures drop.

There has been and continues to be considerable controversy on the causation and function of stereotypies in farm animals. Stereotypic behaviour can be defined as those behaviours that consist of morphologically identical movements that are regularly repeated, have no obvious function or are unusual in the context of their performance (Cronin *et al.*, 1986). Examples of these behaviours are bar-biting (Fig. 2.2), sham chewing, head-weaving and excessive drinking.

It is useful to review discussions of a number of authors on this topic to highlight the current state of knowledge on the causation and function of stereotypies. Let us consider as an example the stereotypy of excessive chain manipulation by sows. Both restraint on tethers and food restriction contribute to the development of this stereotypy in sows; however, sows housed in groups and restrictively fed also show persistent chain-pulling (chains were provided in the pen) (Terlouw *et al.*, 1991), indicating the influence of food level on the development of this stereotypy. The development

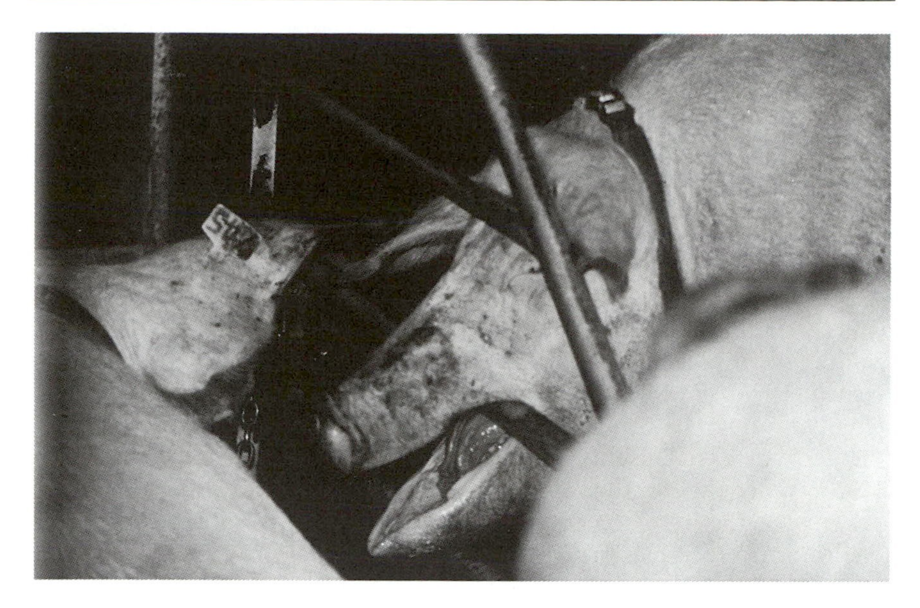

Fig. 2.2. Bar-biting by a pregnant sow housed on a tether in a stall.

of stereotypies may be postulated based on a general model of the control of behaviour (Hughes and Duncan, 1988), in which increasing motivation (e.g. hunger) will trigger appetitive behaviour (e.g. foraging behaviour), which has a positive feedback effect on motivation and directs the animal towards the appropriate goal (e.g. food). The performance of both the appetitive and consummatory (e.g. feeding) behaviours will provide initial feedback on motivation, but, eventually and together with the resulting physiological changes (energy intake), will cause negative feedback, which will switch off the motivational state. In the above example of a stereotypy in restrictively fed sows, the stereotypy may have developed due to the combined effect of a strongly motivated behaviour (e.g. foraging) and an environment that does not allow sufficient expression of the consummatory behaviour (e.g. feeding) to switch off motivation. Activities such as appetitive foraging can be positively reinforcing (rewarding) for the animal and thus, in situations where there is insufficient negative feedback from the consummatory behaviour and its functional components (e.g. food), these appetitive sequences may persist and develop into stereotypies (Terlouw *et al.*, 1991).

Unavoidable fear or stress and barren and restrictive environments have also been implicated in the development of stereotypies. Mason (1991) refers to examples of body-rocking in mentally handicapped patients when distressed and where the incidence of stereotypies increases with increasing confinement. Cooper and Nicol (1991) have proposed that some forms of stereotypies reduce responses to aversion by affecting the animal's perception of the situation. Thus it is clear that different forms of stereotypies may have different causes, such as frustration, stress and lack of control and stimulation; however, our understanding of the motivational basis of stereotypies is poor.

A similar controversy exists in relation to the function of stereotypies. Based on early evidence of associations between stereotypies and physiological signs of coping, such as reduced corticosteroid concentrations, reduced adrenal-gland weights and reduced ulceration, there is a view that stereotypies may be a coping response. However, more recent studies and reinterpretation of some of the early evidence question this general coping hypothesis for at least some forms of stereotypic behaviours (Mason, 1991; Rushen, 1993). Furthermore, while some evidence exists to indicate that stereotypies may be coping mechanisms in the short term, it is unknown whether they exert benefits in the long term. It is a sobering thought that irrespective of the function of stereotypies, the existence of a stereotypy is indicative at the least of a past problem for the animal in coping with its conditions. Stereotypies that result in physical damage or illness to the animal (e.g. the development of lesions in stall-housed sows that persistently rub their tail roots from side to side against stall fittings (Ewbank, 1978) or wind-sucking in horses where persistent wind-sucking can lead to gastrointestinal catarrh and colic (Fraser and Broom, 1990)) have obvious and immediate implications for the welfare of farm animals.

Behavioural change in which there is deviation in either the pattern, frequency or context of the behaviour from that which is generally expected and which results in adverse effects on the morbidity or mortality of the individual or others clearly has welfare implications. For example, tail-biting in pigs, for which the cause(s) is poorly understood, results in restlessness, poor growth and possible paralysis and mortality due to infections in the recipients (van Putten, 1969). Similarly chewing of wood, soil, hair or faeces has serious welfare implications for the animal, through the ingestion of this material.

There is a view that animal welfare is only impaired if the animal is experiencing an unpleasant mental state (Dawkins, 1990; Duncan and Petherwick, 1991). An extrapolation of this argument is therefore that, irrespective of the health or injury status of the animal, welfare is considered to be at risk if the animal's mental state is adversely affected. The considerable interest in cognitive ethology, particularly animal consciousness and cognitive processes, by applied ethologists is partly driven by an interest in animal-welfare assessment. Cognitive ethology is concerned with studies of subjective experiences in animals, and Griffin (1992) has attempted to address questions concerning the nature of other species: are they conscious or self-aware, do they plan ahead or intend to deceive? These questions are of obvious relevance to animal welfare; however, our current knowledge and methodologies do not allow us to directly demonstrate the existence of feelings (Nichol, 1995), such as pleasure, displeasure, fear, anger, frustration, etc. Progress in understanding animal self-awareness has been limited by a lack of a clear understanding of what cognitive tasks are associated with self-awareness. With our present knowledge it is difficult to appreciate the way in which an animal with specific cognitive skills is more likely to suffer than one lacking those skills.

There has been and continues to be considerable interest and activity by scientists in studying the preferences of animals for resources, such as space, flooring and a parturition or nest site. The impetus for this interest is the view that choices may indicate the existence of important underlying needs. The preferences of animals for resources can be studied by allowing the animals to choose between resources and preference is measured in either the time the animal spends with the resources or the resource that is selected. The simplest preference study involves allowing the animal to make a choice between two situations in which the resource is varied. For example, Hughes (1975) found that laying hens preferred a spacious cage to a confined cage and that neither time of day nor strain of bird was influential in this choice. Observing animals in complex environments offering a range of activities will also provide details of the animal's preference for habitats and resources.

In an attempt to measure the strength of an animal's choice, scientists have incorporated tasks in which the animal has to expend energy or take risks to gain access to an alternative resource. For example, operant conditioning techniques (see Chapter 5), in which an animal learns to perform a response, such as a lever pressing, to gain access to the alternative resource, have been used to measure the value that the animal puts on the resource. Hutson (1992) found that preparturient sows worked harder, on the basis of lever lifting, to gain food than access to straw. Consumer-demand theory has been used with preference testing to put a value on the animal's choice (Dawkins, 1983); the strength of motivation ('need') for a resource can be measured through the animal's willingness to consume ('work' for) the resource as the 'price' of the resource increases. Thus, by measuring consumption at increasing prices, needs can be classified as necessities where the animal works harder to maintain consumption (called an inelastic demand function) or luxuries where the animal does not maintain consumption by working harder (elastic demand function). Using this approach, needs can be ranked in terms of their demand functions.

Preference or choice testing has been criticized on several grounds. For example, previous experience may affect choice, with animals initially avoiding a resource simply because it is less familiar or novel. Animals may not always choose wisely in terms of what is good for them in the long term (Duncan, 1978; Barnett and Hutson, 1987), and a good example of this is overeating. Furthermore, an animal's short-term choice may reflect its proximate (immediate) needs rather than the animal's ultimate needs or those necessary for survival, growth and reproduction (Lawrence and Illius, 1997). In other words, since motivational strengths will vary markedly over time in the short term, animal choices are likely to correspondingly vary in the short term. For example, since an animal's choice between feed and space will be markedly affected by short-term changes in hunger, the choice of space is more likely immediately after feeding rather than later. Clearly, there is a need to assess animal choice over relatively long periods in order to measure ultimate rather than proximate needs.

Assessing motivation using preference testing offers us the potential to measure important underlying needs, which obviously is a valuable tool in studying animal welfare. The inclusion of developments in this area, with methodologies evaluating the stress responses and their consequences for the animal, is likely to make a significant contribution to the assessment of welfare in the future.

2.5.4. Conclusion on welfare assessment

A valuable, objective and defensible approach in assessing risk to the welfare of an animal with our current knowledge is to measure firstly the behavioural and physiological stress responses of the animal to the environmental change in question. The second step is to measure any consequent biological cost to or reduced fitness of the animal. Thus risks to welfare are assessed in terms of the magnitude of these responses and also the consequences of these behavioural and physiological responses for the animal's ability to grow, reproduce and remain healthy. This is the approach that is utilized in this book.

Investigations in the immediate future are likely to search for evidence of decreases in fitness, and welfare assessments will be made in relative rather than absolute terms. One system or procedure will be compared with another by comparing the effects of the two on the biological responses of the animals and their consequences for the animals.

2.6. THE VALUE OF THE CONCEPT OF STRESS IN ASSESSING ANIMAL WELFARE

The concept of stress as described in this chapter is utilized in this book and the two following examples, one from a natural population in the wild and the other from agriculture, powerfully demonstrate the usefulness of the concept and its value in assessing welfare. The first is an example from Australian biology that demonstrates the impact of stress and its consequences on animal fitness (Lee *et al.*, 1977). The brown marsupial mouse (*Antechinus stuartii*), a small insectivore, has a very synchronized and dramatic life history, the most dramatic feature of which is a total annual mortality of males prior to the birth of the next generation. This mortality is stress-related and the main stressors are altered behaviour patterns, in turn driven, in part, by endocrine changes. As the 2-week breeding season approaches, males become aggressive towards one another and spend considerable time searching for females. As a result of these behaviours, there is an array of physiological changes during the period of mortality, including an increase in total corticosteroid concentrations (Barnett, 1973), a decrease in transcortin concentrations and consequently a sustained increase in free corticosteroids (Bradley *et al.*, 1980). This sustained increase in free corticosteroid concentrations is evidence for a chronic stress response. Some consequences in this species are a loss in body weight and changes in plasma sodium and glucose concentrations (Barnett, 1973), anaemia (Cheal *et al.*, 1976), haemorrhagic ulceration of the digestive tract (Bradley *et al.*, 1980) and a suppression of the immune system, resulting in a variety of pathological states and mortality (Barker *et al.*, 1978).

The second example, albeit less dramatic, is from a series of experiments on pigs in intensive animal production. Pregnant pigs housed in neck-tethers in partial stalls of a design that permits aggressive interactions with neighbouring tethered pigs develop a chronic stress response. Evidence for this chronic stress response in these pigs is a sustained elevation in free cortisol concentrations, compared with concentrations in pigs housed either in other types of individual housing or in group housing with adequate space (Barnett *et al.*, 1985, 1989, 1991). The consequences of this chronic stress response in these tethered animals include changes in plasma glucose and urea concentrations (Barnett *et al.*, 1985); these changes are indicative of an energy-mobilizing mechanism to adapt to the housing design. In addition, there are decreases in both reactivity of the immune system (Barnett *et al.*, 1987) and pregnancy rate (Barnett *et al.*, 1991). These tethered pigs showed marked changes in their social behaviour in that many aggressive interactions with adjacent neighbours were unresolved. Similarly, overcrowding of mature female pigs results in a chronic stress response and depressions in sexual receptivity (Hemsworth *et al.*, 1986c).

One can argue that these two examples of serious welfare risks involve extreme effects and thus are extreme examples. However, these two cases clearly demonstrate, within the concept of stress outlined earlier, the consequences of animals failing to cope with an environmental change. All stressors, whether physical or emotional, are processed by the central nervous system, which in turn integrates this information and dictates which of the behavioural and physiological systems must respond to maintain homoeostasis. This biological response may come at a physiological cost to the animal, such as decreased metabolic efficiency, impaired immunity and loss of the ability to reproduce. Such biological changes and biological costs for the animal clearly enable the interpretation with some considerable degree of confidence that the welfare of these animals was seriously compromised. Our current knowledge may not allow scientists at this time to detect more subtle or less serious risks to welfare.

2.7. CONCLUSIONS

We have not attempted to address the issue of animal rights in this chapter. There has been considerable argument at the community level and among philosophers and scientists about the extent to which humans are entitled to use animals for a variety of purposes, including as companion animals, as objects of hunting or fishing and in agriculture. It is really beyond the scope of this book to adequately address this issue. Nevertheless, current community values do accept the moral right for humans to use animals in agriculture as a source of food but place considerable emphasis on the welfare issues discussed here.

While the general community and, at times, the livestock industries have implicated housing systems and some husbandry practices as potential causes of reduced welfare in farm animals, the influence of the stockperson on the welfare of animals is not necessarily obvious to many people both within and outside agriculture.

In this chapter we have attempted to provide the background information that is important in exploring the role of human–animal interactions in livestock production, particularly with respect to the implications for animal welfare. It will become clear in subsequent chapters, for example, that the welfare of livestock is at serious risk in situations where the stockperson's commitment to the surveillance of, and the attendance to, welfare and production issues is less than optimal. Similarly, animal performance is also at risk in such situations. This book will therefore highlight the integral role that the stockperson plays in determining the welfare of farm animals. Indeed, as shown in Chapter 1 (Section 1.2), in some situations, human factors may be relatively more important to animal welfare and productivity than the housing system *per se*.

In the next chapter, we shall consider human–animal interactions from the point of view of current research. The aim will be to identify some of the fundamental principles underlying the way in which animals react to humans.

Chapter 3

Human–Animal Interactions and Animal Productivity and Welfare

3.1. INTRODUCTION

In most walks of life, humans frequently interact with animals and, in many situations, these interactions are frequent and intense, to the extent that relationships develop between humans and animals. The human–companion-animal relationships that are so common in Western-society households are an excellent example of the intense and close relationships that may develop between humans and animals. Some of the benefits that the relationship offers to both partners are recognized. Obviously, pets are virtually dependent on humans for their care and maintenance; however, there is growing appreciation of the benefits that companion animals may have for human health and behaviour – for instance, the use of companion animals as guides for the blind, to enrich the lives of long-term patients and even to reduce the risk factors associated with heart disease (Edney, 1992). Furthermore, there is evidence that pet ownership may improve social competence in children and the presence of pets may improve the efficacy of therapy for mentally disturbed people (Edney, 1992). In contrast, there is relatively little appreciation, both within the general community and, to some extent, within agriculture, of the implications of human–animal relationships in agriculture.

In modern livestock production, particularly in intensive-production systems, similar human–animal relationships develop, since farm animals receive frequent and, at times, close human contact. Even though there may be considerable automation with these production systems, stockpeople are required to regularly monitor animals and their conditions and impose routine husbandry procedures. Consequently, the amount of human contact that these animals receive is considerable. For example, a stockperson in modern meat-chicken units may manage 80,000–100,000 birds at a time and, although the stockperson may not physically interact with his or her animals, the stockperson will often be in close visual contact with most of the birds as often as six times daily, during routine inspection of birds and their condition. Lactating dairy cows are intensely handled twice daily during lactation and a stockperson may handle 100–200 cows twice daily at the time of milking. Similarly, breeding female pigs may be handled frequently around mating and parturition, while breeding boars may be handled at least daily during their entire breeding life. Handling of pigs and dairy cattle generally involves tactile interaction by the stockperson, which may be either negative in nature, such as a push or hit, or positive in nature, such as a pat or stroke. Thus, even though the number of animals managed by each stockperson has substantially increased in modern production systems, farm animals in these systems receive frequent and, at times, close human contact.

Many people in both the farming and general community recognize that early experiences with humans may have effects on the subsequent behavioural responses of animals to humans. Popular reports on the effects of early experience with humans, such as those in the studies by Konrad Lorenz (1970) on hand-reared birds, have probably contributed to this general appreciation of the effects of early human contact. However, what is surprising for many is the profound effects of not only this early experience with humans, but also the effects of regular human contact on the long-term behaviour and physiology of the animals. There is good evidence, based on handling studies in the laboratory and studies examining fear–productivity relationships in agriculture, that human–animal interactions may markedly affect the productivity of farm animals. Since fear and stress appear to be implicated in the effects of human–animal interactions on productivity, these interactions between humans and farm animals may also have implications for the welfare of these animals. This chapter examines these effects of human–animal interactions on the productivity and welfare of farm animals. This chapter also forms an introduction to Chapters 4 and 5, in which the development of these human–animal interactions in agriculture is explored.

3.2. HUMAN–ANIMAL INTERACTIONS AND THE HUMAN–ANIMAL RELATIONSHIP

The conceptual framework used to consider human–animal relationships in this book is based on that described by Estep and Hetts (1992), which in turn was derived from the views of Hinde (1970) on social relationships. Social structures can be considered to be built up of relationships, and relationships are built up of interactions. Thus human–animal relationships can be considered to be constructed from a series of interactions between humans and animals. These interactions between humans and animals may be tactile, visual, olfactory, gustatory and auditory and the nature of these interactions may be positive, neutral or negative. For example, fear-provoking interactions, such as the sudden unexpected appearance of a human or a human looming over an animal, may be negative for the animal, while painful interactions, such as a hit by a human, are obviously negative to animals. As will be shown later in this chapter and in Chapter 5, it is the nature and the number of these human interactions that markedly determine the quality of the human–animal relationship for farm animals.

The quality of this relationship from the perspective of the animal can be assessed by measuring the behavioural response of the animal to the handler or to humans. In other words, the behaviour of the animal in the presence of the handler or other humans (e.g. the approach and avoidance responses of the animal to the human) will provide information on the quality of the human–animal relationship for the animal.

3.3. BEHAVIOURAL RESPONSES OF DOMESTIC ANIMALS TO HUMANS

There are marked between-species and within-species differences in the behavioural responses of animals to humans. For example, the flight distance from humans, which generally is defined and measured as the distance at which an animal withdraws or escapes as a human approaches (Hediger, 1964), varies markedly both between and within farm-animal species. Murphey *et al.* (1981) reported marked differences in the flight distance of *Bos indicus* and *Bos taurus* breeds of cattle from humans and Hearnshaw *et al.* (1979) reported marked differences in the behaviour of crossbred Brahman cattle and British breeds to handling. Indeed, the latter authors reported that the behavioural response to restraint in a squeeze shute (or stall) in the close presence of humans, often referred to as temperament, is moderately heritable in *B. indicus* cattle. Furthermore, the flight distance of extensively grazed farm animals is generally reported to be greater than that of intensively managed farm animals, perhaps as a consequence of fewer interactions with humans in the extensive situations. For example, there are reports of flight distances of 6–11 m for extensively grazed or rangeland sheep and 31 m for extensively grazed beef cattle, in comparison with 2–8 m for feedlot beef cattle and 0–7 m for dairy cattle (Grandin, 1980, 1993; Hutson, 1982; Purcell *et al.*, 1988).

These differences in the avoidance responses of animals to humans may, in part, reflect inherent species differences in their fear of unfamiliar stimuli (neophobia). Selection for neophobia is more likely to affect the general fearfulness of the naïve animal, rather than influencing specific responses to specific novel stimuli, a scenario expressed in the behaviour of both wild and domestic animal populations (Price, 1984). Inherent species differences in neophobia will affect the initial responses of naïve animals to novel stimuli, such as humans; however, over time, experience with humans should modify these responses to the extent that these responses become stimulus-specific. Murphy and Duncan (1977, 1978) studied two stocks of chickens, termed 'flighty' and 'docile', on the basis of their behavioural responses to humans, and found that early handling affected the behavioural responses of these two stocks of birds to humans, with the docile birds showing a more rapid reduction in their withdrawal responses to humans with regular exposure to humans than the flighty birds. These stock differences may be stimulus-specific, since observations indicated that the docile birds did not necessarily show fewer withdrawal responses to novel stimuli, such as a mechanical scraper and an inflating balloon, than the flighty birds (Murphy, 1976).

Further evidence that the handling effects on the behavioural response of animals to humans may be specific to humans and not generalized to a range of fear-provoking stimuli is a series of studies by Jones and colleagues (Jones *et al.*, 1991; Jones and Waddington, 1992). These scientists examined the effects of regular handling on the behavioural responses of quail and domestic chickens to novel stimuli (such as a blue light) and humans and found that handling predominantly affected the responses of birds to humans, rather than to the novel stimuli. Handled birds showed less avoidance of humans but their responses to novel stimuli were unaffected. These data indicate that experience with humans result in stimulus-specific effects rather than effects on general fearfulness. In contrast, Lyons *et al.* (1988) reported that early human contact not only affected the behavioural responses of goats to humans but also affected the behavioural responses of goats to a range of novel stimuli; in comparison with dam-reared goats, those reared by humans showed increased approach to and less avoidance of a number of stimuli, including humans. However, human involvement was involved in testing the responses of these goats to novel stimuli and thus their response to humans at the time of testing may have influenced their responses to novel stimuli. In fact, many of the studies that have examined the behavioural responses of animals to novel stimuli have used testing procedures that involved the close presence of humans. Furthermore, in the study by Lyons *et al.* (1988) and, indeed, in a number of other studies, early weaning also occurred in the human-rearing treatment, and this early stressor may have affected the subsequent behaviour of the human-reared goats to novelty in a similar manner to that observed in the so-called 'early-handling studies'; in many of these studies, an early stressor(s) was shown to have marked effects, through its effects on developmental processes, on the animal's subsequent behavioural and physiological responses to novelty (Schaefer, 1968).

Over time, young domesticated animals that may have had limited experience with humans may habituate to the presence of humans and thus may perceive humans as part of the environment without any particular significance. Habituation will occur over time as the animal's fear of humans is gradually reduced by repeated exposure to humans in a neutral context; that is, the human's presence has neither rewarding nor punishing elements. Even wild strains that are highly fearful of humans will habituate to humans. Galef

(1970) tested effects of several rearing experiences on the behavioural response of wild Norway rats to handling by humans. Second- and third-generation rats, which were housed in a laboratory, were reared by either wild or domestic rats, reared with either wild or domestic littermates and received either regular handling or no handling from 10 to 23 days. At weaning, at 23 days of age, the behavioural response of these animals to handling was observed. Only those animals that had had physical contact with humans (handling) showed minimal withdrawal responses to capture. Wild-caught deer and deer bred in captivity habituate rapidly to the presence of humans and flight distance from humans reduces to 30 m or less (Matthews, 1993).

Some domestic animals, such as farm and laboratory animals, and, indeed, some pets, such as aviary birds housed in groups, which may receive limited human contact, may perceive humans as predators. Selection for increased docility in the presence of humans has accompanied domestication; however, based on their withdrawal responses to humans, domestic animals may still find human contact aversive and thus perceive humans as predators rather than benevolent caretakers. *Bos indicus* cattle extensively grazed with infrequent human contact display extreme avoidance responses to restraint and human presence, including at times displaying tonic immobility or a catatonic-like state during restraint (Grandin, 1980), which are indicative of antipredator responses. Caine (1992) has challenged the widely held view that a captive animal's behaviour can habituate to the presence of human observers. Her data suggest that the presence of observers, even for captive animals that have received considerable human contact, may result in antipredator behaviour in these animals, often masking or confounding the behaviour under study.

Domestic animals in situations in which they frequently interact with humans may, through conditioning, associate humans with rewarding and punishing events that occur at the time of these interactions and thus conditioned responses to humans may develop. Studies examining the effects of a range of handling treatments on the behaviour of pigs (see Section 3.5.1) indicate that conditioned approach–avoidance responses develop as a consequence of associations between the stockperson and aversive and rewarding elements of the handling bouts. Pigs that were slapped or shocked with a battery-operated prodder whenever they approached or failed to avoid the experimenter in daily handling bouts of 15–30 s learned to associate the presence of the handler with the punishment of the handling bouts. In contrast, pigs that received pats or strokes during brief daily handling bouts subsequently showed increased approach to humans. Furthermore, there is evidence that pigs may associate the rewarding experience of feeding with the handler and that this conditioning results in pigs being less fearful of humans (Hemsworth *et al.*, 1996d). Although there is some controversy over the mechanism by which avoidance behaviour becomes conditioned by punishment (Walker, 1987), it is well established that animals learn to avoid conditioned stimuli that are paired with aversive events. Thus, through conditioning, the behavioural responses of farm animals to humans may be regulated by the nature of the experiences occurring around the time of interactions with humans.

Animals may also perceive humans as social partners and this is likely in situations where young animals form a long-lasting bond or attachment to humans at an early age. The extensive studies by John Paul Scott (see Scott, 1992) on the effects of early human contact on the socialization of dogs to humans elegantly demonstrate the long-term effects of early human interaction; human contact, as little as a few minutes of daily visual contact with humans or just two 20-min periods of visual contact with humans, from the age of 3 to 8 weeks of age will have profound effects on the subsequent behavioural responses of dogs to humans. Dogs that received this amount of human contact early in life show reduced fear responses to humans in adulthood in comparison with control animals that received minimal human contact early in life.

3.4. FEAR OF HUMANS BY DOMESTIC ANIMALS

The phrase 'fear responses to humans' is a useful and convenient term to describe the behavioural responses of animals to humans. These responses may include escape–avoidance responses. Approach responses to humans occur when fear wanes and the animal commences to explore and investigate the human stimulus, and thus these approach responses can be usefully considered as an inverse measure of the level of fear of humans. Fear, as with other states of motivation, such as sexual motivation, hunger and thirst, cannot be measured or observed directly, but can only be inferred by observing the behaviour of the animal. However, unlike these other states, this limitation in its measurement has contributed to the considerable controversy that exists with the definition of fear, its use by scientists and its measurement (see Hinde, 1970; Murphy, 1978) and therefore it is useful to briefly consider these aspects now.

3.4.1. Definition

Fear can be viewed as an intervening variable, linked, on the one hand, to a range of stimuli that may pose some risk or danger to the welfare of the animal and, on the other hand, a series of responses, both behavioural and physiological, by the animal, which enable it to respond appropriately to this source of danger. Gray (1987) defined fear as a form of emotional reaction to a stimulus that the animal works to terminate, escape from or avoid, and McFarland (1981) considered fear as a motivational state that is aroused by certain specific stimuli and which normally gives rise to defensive behaviour or escape.

Farm animals in the presence of humans commonly display behavioural patterns that can be labelled fear responses, such as withdrawal from or avoidance of humans, as well as immobility responses, such as freezing or crouching, in the close presence of humans (Hemsworth and Barnett, 1987; Jones, 1987; Mills and Faure, 1990). It is reasonable to label these responses as fear responses, since it is generally accepted that fear responses function to protect the animal from harmful stimuli. What is surprising when studying farm animals is the magnitude of these responses, given that these animals have been domesticated over many generations and that there is generally substantial contact between humans and farm animals in modern animal production systems. As Hale (1969) mentions, one of the main behavioural traits that facilitate the domestication process is a short flight distance from humans or low fear levels of humans.

The view that fear is an intervening variable is a useful construct or supposed concept that enables the study of the proximate or immediate causes and ontogeny of these behavioural responses to humans. An understanding of the causal factors may provide the opportunity to reduce fear responses to humans in farm animals, and therefore the main factors that have been identified that regulate these responses in farm animals are considered in detail in Section 3.5.

3.4.2. Behavioural and physiological responses to humans

The behavioural and physiological responses that fearful animals may display in the presence of humans have been described in detail by several authors; however, it is useful to consider these responses since it is these that have important implications for the productivity and welfare of farm animals. These biological responses are the stress responses considered earlier in Chapter 2, Section 2.4, and, since they are the principal means available to the animal in its attempts to cope with a stressor, these stress responses are again reviewed, but briefly and in the context of exposure of the animal to a fear-provoking stimulus.

In the close presence of humans, there are basically three types of interrelated biological responses that are available to an animal that is highly fearful of humans: behavioural responses, responses of the autonomic nervous system and responses of the neuroendocrine system. The central nervous system integrates these responses and it is these responses that provide the principal resources that the animal utilizes in its attempts to cope with the stressor. The close presence of a human is likely to initially initiate a series of adaptive or coping responses, often called the 'fight-or-flight responses' or the 'emergency reaction' (Cannon, 1914), which may include escape or avoidance responses, as well as autonomic responses and neuroendocrine responses, resulting in elevation of catecholamines (e.g. adrenaline), heart rate, blood pressure and body temperature, which prepare the animal for these behavioural responses. Thus these autonomic and neuroendrocrine responses function to mobilize the body's reserves for an appropriate immediate reaction to this challenge.

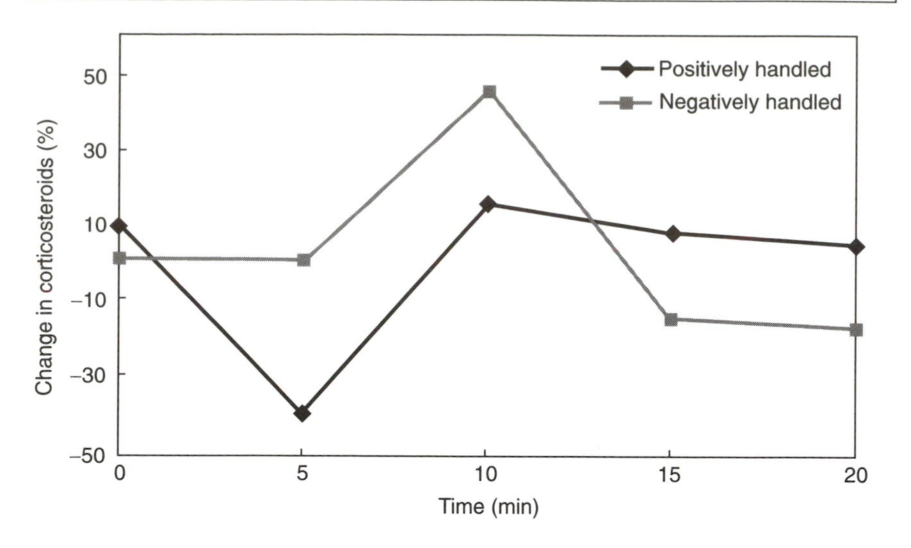

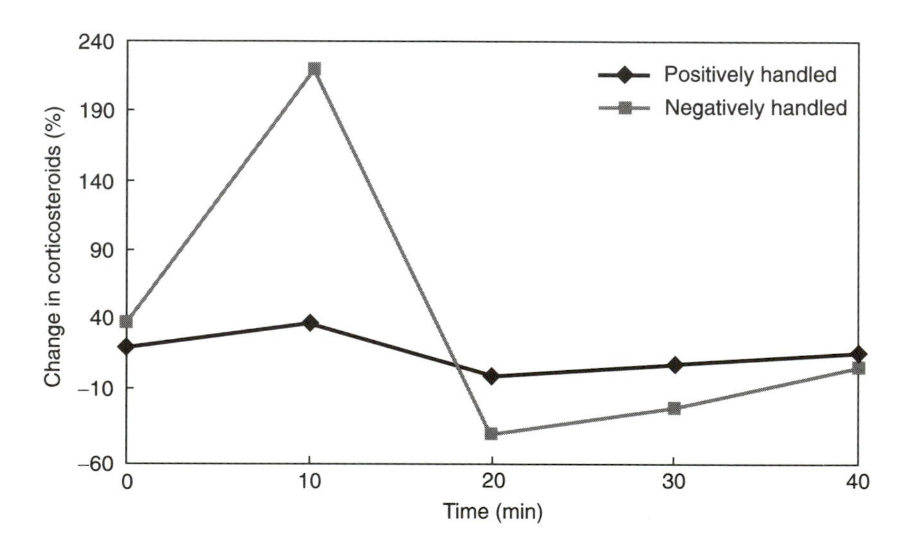

Fig. 3.1. Percentage change in plasma corticosteroid concentrations in positively and negatively handled cows (top figure) and pigs (bottom figure) after a 2-min exposure to a human (data from K. Breuer, unpublished data, and Hemsworth *et al.*, 1981a).

These 'fight-or-flight responses' last for only a short period and, if the stressor (human) is not removed, a second series of events occurs. This is the short-term or acute stress response and is corticosteroid-dependent. Studies on a number of farm animals have shown that highly fearful animals, assessed on the basis of their behavioural responses to humans, will show a greater short-term increase in plasma corticosteroid concentrations within about 15 min of brief exposure to humans than animals that are less fearful of humans (Fig. 3.1). This acute response has the major function of providing glucose from non-carbohydrate sources (particularly protein from muscle) for the required increased metabolic performance necessary to respond to the stressor. If the stressor is removed, this physiological state will disappear with possibly no real ill effects on the animal, apart from a depletion of energy reserves.

The behavioural and physiological responses of the two stocks of birds studied by Murphy and Duncan (1977, 1978), termed 'flighty' and 'docile' on the basis of their behavioural responses to humans, were monitored by Jones *et al.* (1981). Orientation, withdrawal and heart rate were measured as a human slowly approached. The behavioural and physiological changes had similar time courses and the differences observed reflected the original classification of the two strains; when the human was close, the flighty strain showed greater withdrawal and increased heart rate. It is of interest that, when the human was first observed at a distance, the docile birds showed evidence of higher fear, perhaps because the human was not recognized as a human at this distance. As mentioned earlier, these stock differences may be stimulus-specific, since the docile birds do not necessarily show fewer withdrawal responses to novel stimuli, such as a mechanical scraper and an inflating balloon, than the flighty birds (Murphy, 1976).

If the stressor continues (i.e. these short-term responses are ineffective in enabling the fearful animal to avoid or alleviate the challenge of the close presence of the human), the response continues to the third series of events, which is the long-term or chronic stress response. This response is also corticosteroid-dependent and comes at a physiological cost to the animal; prolonged activation of the hypothalamus–pituitary–adrenal axis results in decreased metabolic efficiency and thus growth performance, impaired immunity and reduced reproductive performance. Research, predominantly on pigs, has shown that high levels of fear of humans can result in a chronic stress response, which, in turn, can markedly reduce the growth and reproductive performance of the animal (Section 3.6.1).

3.4.3. Measurement of fear of humans

A number of studies have assessed fear levels by measuring the amount of avoidance of the stimulus or conversely the amount of approach to the stimulus in standard testing situations. The rationale behind these assessments is that, while there may be a number of behavioural patterns available to the animal that may be equally effective in the fear-provoking situation, the amount of avoidance or conversely approach provides an integrated measure of the fear levels, without having to make judgements about the relative significance of specific behavioural patterns. For example, a decision that freezing is indicative of higher or lower levels of fear than orientation away from the stimulus or vigorous escape from the stimulus is not required. Acute physiological stress responses, such as rises in plasma corticosteroid concentrations and heart rate, have also been used to assess fear, often in conjunction with behavioural measures.

In studying the behavioural responses of farm animals to humans, we have adopted a functional view (Hemsworth *et al.*, 1993) and, in studies with pigs and cattle, we have used the approach behaviour of the animal to a stationary experimenter in a standard arena to assess the animals' fear of humans. In these tests, although the degree of novelty of the test arena is reduced because of the similarity of the arena to the animals' home pen, animals introduced into this new environment will be motivated to explore and familiarize themselves with the environment once the initial fear responses have waned. Therefore, although the animals may be motivated to both avoid and explore the arena and the human stimulus, the animal's fear of humans will have a major influence on the animal's approach to the human stimulus.

There is evidence for this interpretation. A significant negative association between the magnitude of the corticosteroid response of pigs exposed to an experimenter in their home pen and the approach behaviour of pigs to a stationary experimenter in the standard test supports our behavioural assessment of fear (Hemsworth and Barnett, 1987). Similarly, Lyons *et al.* (1988) found that the dam-reared goats, which showed less approach to and greater avoidance of humans, had a higher corticosteroid response to human presence than human-reared goats. Furthermore, the imposition of handling treatments designed to differentially affect the pigs' fear of humans produced the expected variations in the approach behaviour of pigs to a stationary experimenter in the standard test (Hemsworth *et al.*, 1981a, 1986b, 1987; Gonyou *et al.*, 1986; Hemsworth and Barnett, 1991).

In studies with poultry, since birds show little locomotion in a novel arena in the short term, the avoidance responses of birds to an approaching human have often been used to assess fear levels (Fig. 3.2). Orientation away and withdrawal from the approaching human have been equated with high fear levels. Handling treatments intuitively expected to reduce the fear of humans by poultry have reduced the incidence of orientation away from the experimenter and reduced the withdrawal from the experimenter in several standard tests measuring the avoidance of humans (Jones and Faure, 1981; Barnett *et al.*, 1992; Jones and Waddington, 1993; Hemsworth *et al.*, 1994c, 1996c).

As mentioned earlier in this section, there are several criticisms of the use of behavioural responses as measures of fear levels. Authors have criticized this approach on the grounds that the responses are not specific to the fear state or motivation and that other motivational states may confound the assessment. However, this is also true when attempting to assess other motivational states. For example, mounting may occur in pigs in both sexual and aggressive encounters. Similarly, a lack of sexual activity by a male in a sexual setting may not be a consequence of reduced sexual motivation

Fig. 3.2. The avoidance behaviour of poultry to humans has been used to assess levels of fear of humans. This photograph shows an experimenter filming the withdrawal responses of chickens as he moves through a broiler-chicken unit in a standard manner.

at the time, but may be a consequence of another motivational state, such as motivation to feed when hungry or fear of the close presence of a predator, inhibiting sexual motivation. As with fear, measurement of these states can also be difficult, relying on inference from observation of the responses. Studies under controlled conditions, in which conflicting motivations are eliminated or controlled, enable the study of the proximate causes and ontogeny of these behavioural responses.

Since a number of behavioural patterns may be available to the animal in a fear-provoking situation, and since a number of these patterns may be equally effective for the animal in avoiding the danger, a number of authors have also argued that there is little scientific basis for ranking these behavioural patterns in terms of fear levels (Murphy, 1978). For instance, is freezing indicative of higher or lower levels of fear than orientation away from the stimulus or vigorous escape from the stimulus? Indeed, this particular problem is one of the main reasons that a functional approach, which does not rely on judging particular patterns of behaviour, has been used to measure fear; a functional approach to the assessment of fear is based on the amount of avoidance of the stimulus or conversely the amount of approach to the stimulus in standard testing situations.

3.5. HUMAN CONTACT AFFECTING THE BEHAVIOURAL RESPONSES OF FARM ANIMALS TO HUMANS

The stimuli from humans that animals respond to when interacting are tactile, visual, olfactory, gustatory and auditory, and these behaviours may influence the immediate behaviour of the animals, as well as the subsequent behavioural responses of the animals to humans. Handling studies on farm animals are useful in identifying the type and nature of human contact that may regulate the behavioural responses of farm animals to humans, and some of the main studies in this literature are considered in this section.

3.5.1. Tactile contact by humans

Handling studies, predominantly with pigs and cattle, indicate that the behavioural response of farm animals to humans is particularly affected by tactile interactions from humans.

Research has shown that pigs are very sensitive to brief tactile interactions from humans. Negative tactile interactions imposed briefly but regularly will produce high levels of fear of humans in pigs. For example, aversive handling treatments, involving brief shocks with a battery-operated prodder or brief slaps whenever the animal approached or failed to avoid the experimenter, daily imposed for only 15–30 s, consistently resulted in pigs showing less approach to humans when subsequently tested with a stationary experimenter in an unfamiliar arena (Hemsworth *et al.*, 1981a, 1986b, 1987; Gonyou *et al.*, 1986; Paterson and Pearce, 1989; Pearce *et al.*, 1989; Hemsworth and Barnett, 1991). In contrast, positive handling treatments, involving pats or strokes whenever pigs approached an experimenter, resulted in low fear levels. Furthermore, pigs handled in a positive manner had a lower acute cortisol response when briefly exposed to an experimenter in their home pens than pigs handled aversively (Hemsworth *et al.*, 1981a, 1986b, 1987). In general, pigs receiving minimal human contact were intermediate in their fear responses. Tanida *et al.* (1995) also found that pigs that were regularly stroked physically interacted with humans more quickly and spent more time interacting with humans than pigs that were not handled.

Handling studies with dairy cows have shown that negative tactile inter-

Fig. 3.3. Handling treatments involving initial passive interactions by the handler, followed by patting and talking on approach by cattle, will reduce fear responses to humans by cattle (Hemsworth *et al.*, 1996c).

actions from handlers will increase their fear of humans. Flight distance in lactating heifers was increased when moderate or forceful slaps were briefly imposed before and after milking when animals failed to avoid the close approach of the handler (Breuer *et al.*, 1997). Similarly, de Passille *et al.* (1996) reported that calves briefly restrained by nose tongs or shocked or threatened with a battery-operated cattle prodder over 7 days spent less time near and were slower to approach and interact with the handler than those that were either briefly petted and fed by the handler or experienced visual contact with the handler.

Handling studies on cattle, sheep and goats that have attempted to reduce levels of fear of humans have generally involved the imposition of positive tactile interactions (for example, Boissy and Bouissou, 1988; Lyons, 1989; Hargreaves and Hutson, 1990; Mateo *et al.*, 1991; Boivin *et al.*, 1992; Hemsworth *et al.*, 1996c) (Fig. 3.3). These studies have generally found that handled animals displayed less avoidance of humans in a range of testing situations designed to assess the behavioural response to humans. In addition to effects on the behavioural response to humans, handled animals in the studies by Boissy and Bouissou (1988), Lyons (1989) and Hargreaves and Hutson (1990) showed lower heart rate and plasma-cortisol responses in a range of situations involving varying amounts of human contact than those that had received less human contact. The handling treatments in these studies with cattle and goats were imposed during rearing, while those on sheep were conducted in adulthood. In addition to presumably positive tactile interactions, such as pats, strokes and fondling, many of these studies involved the experimenter quietly speaking to the animals. Thus auditory and visual interactions were involved in the handling treatments. Furthermore, some of the handling treatments were associated with the provision of food in an attempt to attract the animal to the experimenter. Therefore, as has been shown in pigs (Hemsworth *et al.*, 1996d), the reduced fear responses to humans in some of these studies may have been associated with the reward of food rather than handling *per se*. Studies with horses have shown that active handling early in life, such as holding and rubbing the foal, results in animals being easier to handle subsequently (Waring, 1983).

Most of the handling studies on poultry have utilized handling treatments involving stroking and/or carrying birds, and these have generally been imposed on young chickens of both broiler (meat) and egg-production strains. Although some of the tactile components of the handling treatments may contain negative elements such as stroking birds, many of these treatments often resulted in birds displaying reduced avoidance of humans (Hughes and Black, 1976; Murphy and Duncan, 1978; Jones and Faure, 1981). A number of authors have proposed that the main factor responsible for reduction in fear observed in these studies is habituation of the bird's fear responses over time with repeated exposure to humans. As with some of the studies on other farm animals, the handling treatments often involved the experimenter quietly speaking to the animals and providing food for the animals at the time of the handling bout. Hughes and Black (1976) and Murphy and Duncan (1977) found little or no effect on avoidance behaviour of birds with regard to humans if handling was imposed on adult birds, while Barnett *et al.* (1994) found that increased human contact, predominantly involving visual contact, reduced the subsequent avoidance responses of adult laying hens.

A series of studies have been conducted on the effects of handling, during both the pre- and postweaning periods, on farmed silver foxes (Pedersen and Jeppesen, 1990; Pedersen, 1993). Handling, which involved fondling and talking to the animals, resulted in less withdrawal from the approaching experimenter.

The literature on early handling of rodents is relevant to the discussion on the effects of early handling of farm animals. This literature is very extensive and basically consists of two types of studies, those termed 'handling studies', which involve brief removal of preweaned animals from their home cages, and those termed 'gentling studies', which involve brief stroking of postweaned animals. Although the results have often been contradictory, these treatments have at times resulted in increased growth and accelerated development, reduced activity and defecation in an open-field test, improved performance in learning tasks and physiological stress responses of lower magnitude to subsequent stressors (Dewsbury, 1992). These results have often been interpreted as a consequence of either direct stimulation or acute stress advancing the rate of development of some behavioural and physiological processes. Thus early handling effects may not necessarily be a consequence of handling *per se* but may be a consequence of acute stress early in life (Schaefer, 1968). As mentioned earlier (Section 3.3), this also has implication when interpreting the effects of early-handling studies that may have been confounded by early weaning.

Thus, for many farm animals, the tactile interactions by humans are important determinants of the animals' fear of humans. However, there is evidence that other forms of human contact may affect the behavioural responses of farm animals to humans.

3.5.2. Visual and auditory contact with humans

Obviously the handling studies on farm animals described in the previous section also involved visual contact with humans and thus visual cues may have contributed to these handling effects. A number of studies, particularly on poultry, have specifically examined the effects of human visual contact on fear.

Evidence with poultry indicates that visual contact with humans is effective in reducing levels of fear of humans. Jones (1993) found that regular treatments involving the experimenter placing his/her hand either on or in the chicken's cage and allowing birds to observe other birds being handled resulted in reductions in the subsequent avoidance behaviour of young chickens to humans. Furthermore, the treatment involving the experimenter's hand in the chicken's cage was more effective than regular handling, which involved picking up and stroking the bird. It is possible that the latter treatment, which involves active interaction, may contain some aversive elements for the bird such as picking the bird up or stroking the bird. Barnett *et al.* (1994) also found that regular visual contact, involving positive elements, such as slow and deliberate movements, reduced the subsequent avoidance behaviour of mature laying hens in comparison with minimal human contact that at times contained elements of sudden, unexpected human contact (see Table 3.2). Since the unexpected appearance of a stimulus is fear-provoking for most animals (McFarland, 1981), the birds in the study by Barnett *et al.* (1994) may have been sensitized rather than habituated to the repeated but unexpected exposure to humans, with fear levels therefore increasing rather than declining. The observations that laying hens housed in the top tier of multitiered battery cages and the outside rows with narrow corridors are highly fearful of humans have at least partly been interpreted in terms of frequent unexpected close appearance of stockpeople during routine husbandry (Hemsworth and Barnett, 1989).

There is some limited evidence that young pigs may also be affected by visual contact with humans. Humans standing erect or approaching young pigs have been shown to be more threatening to pigs than humans squatting or avoiding pigs (Hemsworth *et al.*, 1986a).

Thus poultry appear to be particularly sensitive to visual contact with humans and, indeed, positive visual contact may be more effective in reducing levels of fear of humans than human tactile contact. Relatively little is known of the negative visual interactions from humans that may elevate fear levels in farm animals; however, rapid speed of movement by humans and sudden and unexpected exposure to the humans may be fear-provoking, particularly for poultry.

3.5.3. Olfactory and auditory contact with humans

Few studies have been conducted to examine olfactory and auditory contact with humans on farm animals. Very limited research with pigs has suggested that masking of human odours may be fear-provoking for pigs. Young pigs showed less approach to a human wearing gloves than to the ungloved human (Hemsworth *et al.*, 1986a). It is possible that human odours, particularly on the hands, are used by pigs in recognition of humans and that masking these odours may create uncertainty or novelty for the animals.

The handling treatments in a number of the handling studies involved auditory contact by the human, particularly to encourage the animal to approach in order to be stroked or patted. Hemsworth *et al.* (1986a) found no differences in the approach behaviour of young pigs to humans using a loud harsh voice or a soft quiet voice. Animals experienced with humans may learn to associate auditory cues, such as a loud harsh voice, with negative behaviours, such as slaps or rapid movements, since auditory and, indeed, other cues, such as olfactory cues, may accompany tactile interactions by humans.

Therefore, in conclusion, tactile and visual interactions by humans are important determinants of the fear responses of many farm animals to humans. Furthermore, limited interaction appears to be influential in determining fear responses to humans, indicating the sensitivity of farm animals to brief human contact, which at times may appear to be innocuous and inoffensive.

3.6. EFFECTS OF FEAR OF HUMANS ON THE PRODUCTIVITY OF FARM ANIMALS

Handling studies on fear and productivity of farm animals demonstrate the adverse effects of fear of humans on animal productivity. The results of these handling studies, together with some observations on fear–productivity relationships in agriculture, are reviewed in this section.

3.6.1. Pigs

Negative tactile interactions imposed briefly but regularly on pigs not only result in high levels fear of humans (see Section 3.5.1), but may also markedly

Table 3.1. The effects of handling treatments on fear, stress physiology and productivity of pigs in six studies.

Experiment and parameters	Positive treatment	Minimal treatment*	Negative treatment
Hemsworth *et al.* (1981a)			
Time to interact with human (s)†	119	–	157
Growth rate (11–22 weeks in g day^{-1})	709	–	669
Cortisol concentrations (ng ml^{-1})‡	2.1	–	3.1
Gonyou *et al.* (1986)			
Time to interact with human (s)†	73	81	147
Growth rate (8–18 weeks, g day^{-1})	897	881	837
Hemsworth *et al.* (1986b)			
Time to interact with human (s)†	48	96	120
Pregnancy rate of gilts (%)	88	57	33
Cortisol concentrations (ng ml^{-1})‡	1.7	1.8	2.4
Hemsworth *et al.* (1987)			
Time to interact with human (s)†	10	92	160
Growth rate (7–13 weeks, g day^{-1})	455	458	404
Cortisol concentrations (ng ml^{-1})‡	1.6	1.7	2.5
Hemsworth and Barnett (1991)			
Time to interact with human (s)†	55	–	165
Growth rate (from 15 kg for 10 weeks in g day^{-1})	656	–	641
Cortisol concentrations (ng ml^{-1})‡	1.5	–	1.1
Hemsworth *et al.* (1996b)			
Time to interact with human (s)†	52	79	145
Growth rate (from 63 kg for 4 weeks in kg day^{-1})	0.97	1.05	0.94
Adrenal weights (g)	3.82	4.03	4.81

* Treatment involving minimal human contact.
† Standard test to assess level of fear of humans by pigs.
‡ Blood samples remotely collected at hourly intervals from 0800 to 1700 h.

reduce the growth and reproductive performance of pigs. A summary of some of the results of handling studies at our laboratory are presented in Table 3.1. Furthermore, observations in the Dutch and Australian pig industries have revealed significant relationships, based on farm averages, between fear of humans and reproductive performance in pigs (Hemsworth *et al.*, 1981b, 1989). The direction of the relationships indicates that reproductive performance was low at farms where breeding females were highly fearful of humans and the magnitude of these relationships indicates that variation in fear of humans accounted for about 20% of the variation in reproductive performance across the study farms. The magnitude of the associations between fear and productivity was remarkably similar in these two on-farm studies (e.g. the correlation coefficients, which estimate the degree of association, between time to approach within 0.5 m of the experimenter and farrowing rate were −0.55 and −0.54, $P < 0.05$; Hemsworth *et al.*, 1981b, 1989). In the second study, as distinct from the first, farms varied substantially in terms of size, housing systems, genetics, nutrition and locality and yet significant fear–productivity relationships were found, which demonstrates the robustness of the fear–productivity relationship in the industry.

Seabrook and Bartle (1992) have also reported depressions in the growth of pigs following aversive handling. In contrast, Paterson and Pearce (1989, 1992) and Pearce *et al.* (1989) found no effects of regular aversive handling on the growth performance and stress physiology of young pigs. There is no obvious explanation for this lack of effects in the studies by Paterson and colleagues; however, differences between studies in the nature, amount and imposition of handling treatments may be responsible for these apparently contradictory results. For example, a behavioural response of animals to an apparently aversive stimulus (e.g. withdrawal from aversive handling by humans) in some situations may be an effective strategy to enable the animals to cope with this stimulus without having to resort to any long-term physiological adjustment. There may also be genetic differences between pigs in their ability to cope with chronic stressors; however, there is little evidence of this in the literature. Barnett *et al.* (1988) found that, although there were differences between two genotypes of pigs in their basal cortisol concentrations, both genotypes exhibited similar pituitary–adrenal axis responses (in terms of percentage increases in cortisol in the long term) to restraint on tethers.

The magnitude of the fear–productivity relationships observed in the pig industry demonstrates that fear of humans can be considered to be a major factor associated with reduced productivity of commercial pigs. The mechanism responsible for the adverse effects of high fear on productivity appears to be a chronic stress response, because, in a number of experiments on pigs, handling treatments that resulted in high fear levels also produced either a sustained elevation in the basal concentrations of the stress hormone cortisol or enlargement of the adrenal glands, together with depressions in growth and reproductive performance (Barnett *et al.*, 1983; Table 3.1). There is considerable evidence in the literature that stress hormones may adversely affect growth and reproductive performance by disrupting protein metabolism and key reproductive endocrine events (Klasing, 1985; Moberg, 1985; Clarke *et al.*, 1992).

3.6.2. Poultry

Handling studies on poultry generally indicate that handling treatments likely to increase the birds' fear of humans may depress the growth performance of chickens. For example, in experiments with young chickens, Gross and Siegel (1979, 1980, 1982) found that birds receiving frequent human contact of an apparent positive nature, such as gentle touching, talking and offering food on the hand, from an early age had improved growth rates and feed efficiency and were more resistant to infection than birds that either received minimal human contact or had been deliberately scared. Deliberate scaring in the third treatment involved shouting and banging on the birds' cages. Although the behavioural response of the birds to humans was not quantified, the authors stated that the handled birds were easier to handle during weighing and blood sampling than the other birds. Other studies in which positive handling was utilized have also shown that handling is associated with increased growth performance of chickens (Thompson, 1976; Gross and Siegel, 1980, 1982; Jones and Hughes, 1981; Collins and Siegel, 1987). In contrast, Reichmann *et al.* (1978) found no effects of handling on the growth performance of either young broiler or layer chickens, whereas Freeman and Manning (1979) suggested that regular handling decreased growth performance in layer chickens. Since handling may vary from positive to negative in nature for birds, variation in the nature of handling between these studies, through effects on fear, may have been responsible for the variation in the effects of handling on growth performance. For example, Buckland *et al.* (1974) demonstrated negative effects of aversive handling (blood sampling by cardiac puncture) on the growth performance of broiler chickens.

A recent handling study at our laboratory on adult poultry also indicates that high fear levels will limit the productivity of poultry. Barnett *et al.* (1994) found that regular visual contact, involving positive elements, such as slow and deliberate movements, which reduced the subsequent avoidance behaviour of mature laying hens, resulted in higher egg production than a treatment that involved minimal human contact (Table 3.2). The authors speculated that the lower productivity of birds in the latter treatment may be a consequence of a chronic stress response, since there was evidence of immunosuppression in these highly fearful birds.

Studies conducted on broiler chickens and laying hens in the field also support the proposition that high levels of fear of humans may limit the productivity of commercial birds. Significant negative relationships, based on farm averages, were found between the level of fear of humans and the productivity of commercial broiler chickens and laying hens. The egg production of laying hens at the farm was inversely related to the level of fear of humans by birds at the farm (Barnett *et al.*, 1992), as was the efficiency of feed conversion of broiler chickens (Hemsworth *et al.*, 1994c, 1996a). For example, avoidance by broiler chickens of the approaching experimenter accounted for 29% of the variation in feed conversion efficiency across 22 commercial farms (Fig. 3.4; based on data from Hemsworth *et al.*, 1994c). Similarly, in an experiment examining the effects of cage position on fear and egg production of laying hens, level of fear of humans was significantly and negatively related to egg production and efficiency of feed conversion (Hemsworth and Barnett, 1989). In observations on the behavioural

Table 3.2. Effects of handling on the productivity and behavioural response of laying hens to humans (from Barnett *et al.*, 1994).

	Handling treatments*	
	Visual	Minimal
Behaviour†		
Times in front of cage	2.1	1.2
Times orientated forward	2.5	1.8
Physiology†		
Corticosterone concentration ($nmol\ l^{-1}$) posthandling	1.94	2.31
Productivity		
Egg production per day	0.89	0.83

* Visual treatment involved regular visual human contact of a positive nature while minimal treatment involved minimal human contact.
† Behavioural response was assessed by observing the withdrawal response of birds in their cages to an approaching experimenter and corticosterone concentrations were measured immediately following handling.

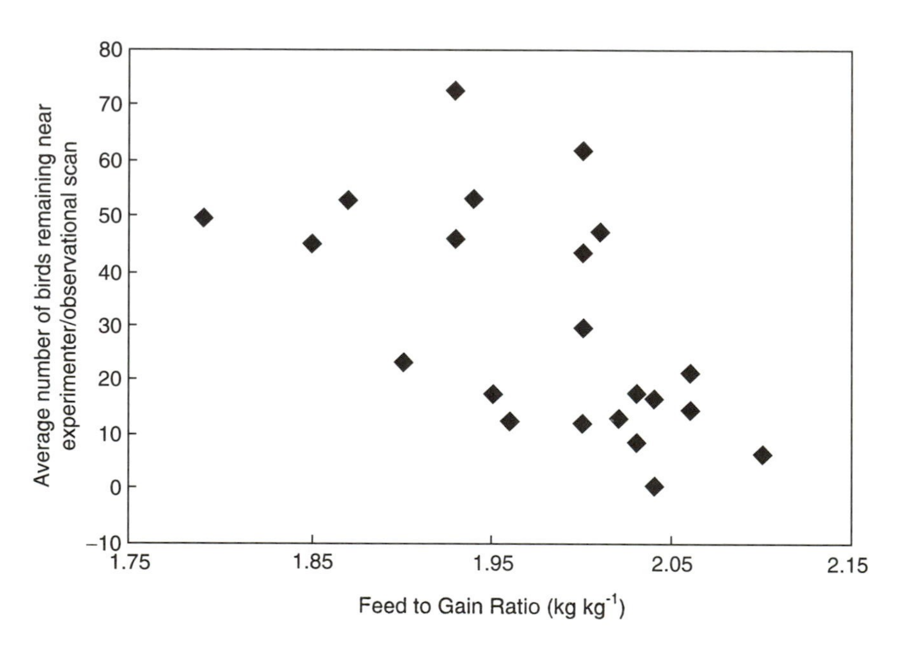

Fig. 3.4. The relationship, based on farm averages, between fear of humans and feed conversion at 22 commercial broiler chicken farms.

response of laying hens to an experimenter, Bredbacka (1988) reported that egg-mass production was lower in hens that showed increased avoidance of humans.

Thus there is evidence that high fear levels may reduce the productivity of poultry. The mechanism(s) responsible is unclear. As seen in fearful pigs, a chronic stress response or even a series of acute stress responses in the presence of humans may be responsible for the depressed productivity in fearful poultry. Support for this suggestion is provided by the known effects of corticosteroids on nitrogen balance, protein catabolism and energy retention or excretion in chickens (Siegel and van Kampen, 1984). Further, exogenous elevations of circulating corticosterone concentrations adversely affected growth rate and efficiency in chickens (Bellamy and Leonard, 1965; Adams, 1968; Bartov *et al.*, 1980; Siegel and van Kampen, 1984; Saadoun *et al.*, 1987). Therefore, regular exposure to stressful stimuli, with the consequent elevations of plasma corticosterone concentrations, might also be expected to impair productivity.

3.6.3. Dairy cattle

There have been relatively few handling studies conducted on dairy cows. Two recent studies suggest that aversive handling may depress the milk yield of cows. Rushen *et al.* (1997) reported that the presence of an aversive handler, who over a 3-day period hit or occasionally used a battery-operated prodder, reduced subsequent short-term milk yield by 10% and increased residual milk (milk not released at a milking) by 71%. Moderate or forceful slaps imposed briefly before or after milking, when animals failed to avoid humans, increased flight distance and tended to reduced milk yield in heifers (6% reduction in yield over 8 weeks; Breuer *et al.*, 1997).

Recent research in the Australian dairy industry examining the relationships, based on farm averages, between level of fear of humans and productivity of dairy cows indicates the existence of significant negative relationships (Hemsworth *et al.*, 1995). Both the approach behaviour of dairy cows to humans and the amount of restlessness displayed by cows in the close proximity of the stockperson during milking were significantly correlated with milk yield of the farm. For example, the correlation coefficient, based on farm averages, between the average time spent within 3 m of the experimenter in a standard test to assess fear of humans and milk yield of the farm was +0.46 ($P < 0.05$). The direction of these correlations indicates that farm productivity was highest where cows showed increased approach to the experimenter in a standard approach test and reduced restlessness during milking. A statistical analysis (regression analysis) to determine the contribution of cow behaviour to the prediction of cow productivity indicated that these two behaviour variables accounted for about 30% of the variation in milk yield across the study farms.

Seabrook (1972a) found that cows in high-yielding herds in Britain tended to be the most willing to approach the milker, to return from pasture and to enter the milking parlour. This report therefore also suggests that milk yield may be at risk when cows are fearful of humans.

It is also of interest that Arave *et al.* (1985) found that dairy calves reared in visual and tactile isolation from other calves produced more milk in adulthood than herd-mates reared either in groups or individually but with visual and tactile contact with calves. In contrast, no such effects were found in a more recent study (Arave *et al.*, 1992). The authors of the earlier study proposed that human-reared calves may have 'imprinted' upon the stockperson and thus may have adapted more easily to the milking procedure, which involves intense human contact. Creel and Albright (1988) rejected this hypothesis on the basis of similar approach behaviour of isolated and control calves to a stationary experimenter. However, they also found that the isolated calves had a shorter flight distance from an experimenter than control calves. Dam-reared goats, which showed increased avoidance of humans, were found to have greater milk ejection impairment than human-reared goats (Lyons, 1989).

In conclusion, there is limited evidence from observations in the dairy industry and handling studies on dairy cows that fear of humans may limit the productivity of commercial dairy cows. These results support those of similar but more extensive studies on pigs and poultry. Further research is required to examine the effects of high fear levels on the productivity of dairy cattle.

3.6.4. Other farm animals

There is very limited evidence that high fear levels may reduce the productivity of other farm animals. This evidence applies generally to situations in which animals have intense or frequent contact with humans and thus in these situations there are opportunities for fear responses to impair productivity.

As referred to in the previous section, dam-reared goats, which showed increased avoidance of humans and lower cortisol responses in the presence of humans, exhibited greater milk-ejection impairment than human-reared goats (Lyons, 1989). The dam-reared goats received considerably less human contact than the human-reared animals.

There is limited evidence that beef cattle which are the most difficult to handle may suffer depressions in their productivity. Fordyce *et al.* (1988) found that beef cattle that were the most active and vocal when restrained in a weighing stall had the most carcass bruising and tended to have tougher meat following slaughter. Although part of the behavioural responses of cattle when restrained in a weighing stall would be responses to restraint and novelty, a component of these responses would be specifically to humans. In studying a similar behavioural response to restraint, Burrow and Dillion (1997) found that the exit speed of beef cattle was negatively correlated with weight gain. This variable of exit speed appears to be highly heritable (Burrow *et al.*, 1988). Although previous human contact for these extensively grazed cattle would have been limited, the contact would have generally been associated with aversive experiences, such as restraint, castration and branding, and thus these observations may reflect neophobia and/or fear of humans.

3.7. CONCLUSIONS: FEAR, PRODUCTIVITY AND WELFARE

In intensive livestock production, there is frequent and often close contact between stockpeople and animals, particularly young animals and breeding animals, and as a consequence of these interactions, many of which are far from superficial, long-term relationships develop between humans and animals. There is evidence from studies both in the industry and in the laboratory that these relationships may exert substantial effects on the behaviour, physiology and productivity of farm animals. Most of the evidence in the literature is from studies in pigs and poultry; however, there is some evidence from recent studies in dairy cattle. It is therefore proposed that the human–animal relationship may have practical implications for farm animals in production systems in which there are close or frequent human–animal interactions.

The precise mechanism(s) responsible for this inverse fear–productivity relationship in some species is unclear; however, based on studies particularly in pigs, it appears that a stress response associated with high fear levels is responsible for limiting productivity. Fear of humans may also have important implications for the welfare of farm animals. The threat to welfare of these fearful animals arises because of injuries that they may sustain in trying to avoid humans during routine inspections and handling, the evidence that these animals when regularly interacting with humans are likely to experience a chronic stress response and, finally, the effects of this chronic stress response on immunosuppression (Hemsworth *et al.*, 1993), which in turn may have serious consequences for the health of the animals. The magnitude of the observed effects on productivity indicates the large potential for improving the productivity and perhaps the welfare of farm animals by identifying and manipulating those human factors which are influential determinants of animal fear in commercial units. The following chapters explore these opportunities.

Chapter 4

Attitudes of Stockpeople

4.1. INTRODUCTION

In Chapter 3, the nature and frequency of human–animal interactions in agriculture were discussed in detail and it was shown that these interactions, which can result in the formation of long-term relationships between humans and animals, may have substantial effects on the behaviour, physiology and productivity of farm animals. The existence of these apparently influential human–animal relationships indicates the potential for improving the productivity and welfare of farm animals by identifying and manipulating the key human factors that regulate these relationships.

Handling studies, particularly with pigs and poultry, have shown that the behaviour of humans may influence the stress physiology and productivity of farm animals and thus the behaviour of stockpeople is a likely candidate implicated in the effects of human–animal interactions on farm-animal productivity and welfare. The psychology literature reveals that the important dispositional factor in predicting human behaviour is attitude. Therefore, in order to improve animal performance and welfare by manipulating stockperson behaviour in agriculture, a thorough understanding of the development of attitudes and their relationship with behaviour is obviously required.

The aim in this chapter is to discuss attitudes and their relationship to behaviour. A limited number of concepts within a theoretical framework will be used to consider the development and maintenance of attitudes and how attitudes can be used to predict human behaviour. This will provide a sound basis for reviewing stockperson behaviours in Chapter 5: how they develop and their role in regulating farm-animal behaviour. Furthermore, the concepts and theoretical framework considered in both Chapters 4 and 5 will be utilized in Chapter 6, in which a model of stockperson–animal interactions, based on research in intensive-farming industries, will be described.

4.2. WHAT ARE ATTITUDES?

The term attitude is widely used in everyday conversation and the media and we all think we understand what it means. Typically, we think of attitudes as opinions – what we think about other people, things and events. In fact, opinions are verbal expressions of attitudes and our describing of attitudes as opinions does not really provide any insight into the nature of attitudes, how they are formed and how they may influence behaviour. In order to understand the nature of attitudes and their relevance to the actions of stockpeople in the workplace, it is important to look more closely at the various meanings of the term.

Eagly and Chaiken (1993) defined attitude as 'a psychological tendency that is expressed by evaluating a particular entity with some degree of favour or disfavour'. There are three key features to this definition: (i) the idea that attitudes are directed at an entity or thing; (ii) the idea that attitude is a tendency or disposition; and (iii) that attitude expresses some positive or negative evaluation. It is important to realize that attitudes cannot be observed directly; we infer people's attitudes from what they say and do.

Characterizing an attitude object is not simple. We may have attitudes towards people, animals, inanimate objects or even ideas. For example, we may have an attitude towards a political ideology, a religion, an individual, a race, a species of animal or a particular animal. As we shall see later in this chapter, to get some insight into expected behaviour, it is very important to be quite specific about the particular attitude object of interest.

The notion of attitude as a tendency reflects the view held by psychologists that our attitudes tend to direct our behaviour or, at least, our intended behaviour (e.g. Ajzen and Fishbein, 1980). For example, if we believe that a particular brand of motor vehicle has outstanding performance and appearance and we like it more than any other brand, we are likely to have a positive attitude towards that car and, when in the market to buy a car, we would tend to buy that particular brand. However, it is easy to see that there may be a whole range of issues, apart from attitude, that we would need to consider before buying the car. These include whether we can afford the car, whether we can sell our current car, what family and friends think of it and what cars friends and neighbours drive. The role of some of these kinds of factors, apart from attitude, in determining stockperson behaviour will be discussed in Chapter 6.

The evaluative nature of attitudes is what distinguishes them from other kinds of verbal expression. A statement that either explicitly or implicitly characterizes something as good or bad, as liked or disliked or even as something to be enjoyed or not enjoyed expresses an evaluation of that object and therefore reflects the underlying attitude. Attitudes, therefore, are favourable or unfavourable; they reflect a tendency for or against, like or dislike, etc.

Historically, psychologists have defined three components to attitude:

cognition, affect and conation (Allport, 1935). Cognition refers to the thoughts that people have about some object. In other words, cognitions are beliefs or subjective facts. They are things which people believe to be true about a person or object. A stockperson may believe that pigs are very difficult to handle and require a lot of effort. This may reflect an underlying negative attitude towards working with pigs. Affect refers to the emotional response that a person has towards some other person or object. The extent to which we like or dislike an object is an example of affective response. A stockperson may express a dislike of pigs or may find them dirty, greedy or smelly. Such expressions would be affective statements reflecting an underlying negative attitude. Finally, conation refers to a tendency to behave in a particular way. A stockperson's intention to avoid contact with pigs or to finish work in the piggery as quickly as possible are examples of conation and may also reflect an underlying negative attitude towards pigs.

There has been a lot of discussion about the three-component concept of attitude. In general, research seems to favour the view that these three components do independently contribute to a person's attitude towards an object (Eagly and Chaiken, 1993). It is important to realize that, while these three components do seem to characterize attitude, they are generally all correlated with each other and all contribute to an underlying evaluative attitude dimension. This means that measuring any one of the components will provide some indication of a person's attitude.

4.3. MEASURING ATTITUDES

Attitudes cannot be measured directly but a person's responses to a series of attitude statements in a questionnaire can be used to infer an underlying attitude. These statements are usually designed to measure one or more of the three components of attitude: the person's beliefs about the object (cognition), emotional response to the object (affect) and behavioural tendency towards the object (conation). Thus, when we assess a person's attitudes for the purpose of predicting subsequent behaviour towards the attitude object, we are actually using a series of verbal responses to predict this behaviour by the person.

4.4. FUNCTIONS OF ATTITUDES

Katz (1960) proposed that attitudes serve four functions. First, attitudes may meet the individual's need to organize experience. Thus a person may express a positive attitude towards a job if he or she has observed others enjoying similar work and if it is felt that it has status and security. A positive attitude to work is a consistent way of reflecting these various aspects of the job. Second, attitudes may help the person to maximize positive experiences and to minimize negative experiences. It is a reflection of the learning processes a person undergoes in the job. Thus negative aspects of a job, such as a history of adverse workplace experiences, poor pay and little opportunity for advancement, may lead to a poor attitude to work. As a consequence, a person will minimize emotional and, perhaps, physical investment in the job. Third, attitudes may serve to protect a person from negative events. The expression of authoritarianism in ethnocentric attitudes (that is, attitudes which express a negative view of other races) is an example of this. An authoritarian person protects him- or herself from the possible threat from an outgroup (a group other than one's own social, ethnic or work group) by assuming a superior stance. Finally, attitudes are an expression of the personality of the individual. Thus a person's attitudes are consistent with self-image. In agriculture, a person who takes pride in being a good stockperson may hold positive attitudes towards husbandry, acquisition of knowledge and work ethic. The identification of these functions of attitudes reflects the belief among psychologists that attitudes have a motivating influence on our behaviour.

In general, although there have been various recent elaborations of these functions of attitudes, Katz's (1960) basic functions of attitudes remain current (Eagly and Chaiken, 1993).

4.5. DEVELOPMENT OF ATTITUDES

Attitudes are generally regarded as learned attributes of a person. They develop initially as part of the socialization of the individual. Young children are dependent primarily upon their parents for information about how they should behave towards others. As a result, children's attitudes come initially from their parents and other close family. Once the child reaches school age, there is an increasing influence of teachers, school friends and television. Stagner (1961) reports a study by Blake and Dennis (1943), who tracked the development of white children's attitudes towards American blacks. Initially, very young children showed no particular attitude towards blacks. As the children developed, they formed more and more negative stereotypes, which were like those of their parents. One important aspect of the development of these attitudes is that they need not be based on direct experience. Children learn to accept what parents say because parents control the rewards and punishment. Stagner (1961) quotes a report from Horowitz and Horowitz (1938) of a second-grade girl as saying: 'Mother doesn't want me to play with coloured children, 'cause they coloured men. Might have pneumonia if you play with them. I play with coloured children sometimes but mother whips me.'

In school-age children and in adults, social pressure to conform in beliefs and behaviour comes from friends, from school- or workmates and from the media. In a classic study, Asch (1956) demonstrated that people would change their beliefs and their behaviour when pressured by others. How likely they were to change and the amount of this change depended on the size of the difference between the person's beliefs and the others' beliefs and the extent to which the person was in a minority. This tendency to change is called conformity behaviour.

4.6. ATTITUDE SYSTEMS

A person's attitudes do not exist independently of his or her other attitudes. If, for example, a person believes that excessive drinking is bad, that person will also have related attitudes towards the cost of alcohol, drinking at lunch-time and so on. In other words, there will be a system of attitudes which are more or less consistent with each other. Not only are these attitudes consistent, but the person will also tend to behave in a way consistent with those attitudes. For example, the person may prefer the lunch-time company of people who don't drink so that he or she will not be under social pressure to drink. While the origin of this tendency for consistency is a matter of some argument (Ajzen, 1988), there is widespread acceptance of it as a well-established characteristic of human behaviour.

One of the most widely accepted theories of attitude consistency is that of Leon Festinger. Festinger (1957) proposed his theory of cognitive dissonance to account for the way in which attitudes influence behaviour, and this theory was later extended by Brehm and Cohen (1962) and Aronson (1969). Dissonance theory proposes that cognitive elements (that is, beliefs or any pieces of knowledge) are dissonant if one element does not follow from the other and are consonant if one element does follow from the other. For example, the statement 'I smoke cigarettes' is dissonant with the belief 'smoking is a health hazard' but is consonant with the belief 'smoking relaxes me'. The theory is very broad – anything that we think about can be regarded as a cognitive element.

Cognitive-dissonance theory has been the focus of a great deal of research. In particular, it has led to a focus on the ways in which people process new information in the context of their existing attitudes. It provides a framework within which the interrelationships between attitudes and behaviours can be understood. This is particularly important when we seek to understand attitude change and to predict behaviour from attitudes.

4.7. ATTITUDES AND BEHAVIOUR

Most people, and psychologists are no exception, attempt to explain a person's behaviour in terms of dispositions. In other words, we believe that people tend to do those things that are consistent with their underlying characteristics. For example, an authoritarian person is likely to treat superiors with deference and unquestioning obedience while treating inferiors with arrogance and detachment. As we have seen, attitudes represent a major class of dispositions. The question arises, therefore, how well do attitudes predict an individual's behaviour? Despite an intensive research programme directed towards clarifying the relationship between attitudes and behaviour, Hovland and his coworkers at Yale University were unable to demonstrate that attitudes were good predictors of behaviour (Hovland *et al.*, 1953). This research involved the assessment of people's attitudes towards objects and the establishment of the relationship of these attitudes to specific behaviours. As we shall see shortly, generic attitudes are not good predictors of behaviour.

The advertising industry is based on the assumption that changing people's beliefs about a product will influence them to buy the product. This is only true in a sense. If a new cigarette is introduced into the market, it is very difficult to determine whether any given individual will buy it. It will depend on whether the person is already a smoker, how loyal he or she is to a current brand, whether or not the person is thinking of giving up smoking, and so on. In fact, the advertising industry relies on a small percentage of consumers changing their attitudes and behaviour; it is an actuarial exercise rather than one of changing a particular individual's attitudes.

A very interesting study which demonstrated the uncertain relationship between attitudes and behaviour was carried out by La Piere in 1934. In this study, all but one of the operators of 250 restaurants and hotels served a visiting Chinese couple when the couple visited unannounced. When La Piere sent questionnaires in the mail to the same proprietors, 92% indicated that they would not be willing to accept Chinese guests. The question which immediately arises is 'Why the discrepancy?' Here, the questionnaire answers were clearly at odds with the actual behaviour of the hotel proprietors. In fact, it is possible to think of many reasons – the proprietors may have wished to avoid a scene with other guests, or may have decided that this particular Chinese couple were respectable and not the kind of people they had in mind when answering the questionnaire.

This illustrates that it is very important to identify the attitude object precisely if attitude is to be used to predict behaviour. In La Piere's (1934) study, for example, the questionnaire would really have had to describe the particular Chinese couple in some detail and also to have characterized the circumstances of their visit, in order to get a reasonably accurate idea of how the proprietors would behave.

It also illustrates that a specific behavioural situation may not provide the ideal context in which to predict behaviour from attitudes. Campbell (1963) asserted that a single behavioural event may not be sufficient to provide an accurate measure of behaviour and that patterns of behaviour should be measured. Fishbein and Ajzen (1974) reported that, when predicting a single behaviour from attitudes toward religion (e.g. regular church attendance), correlations of 0.12–0.15 were obtained, depending on the method used. When multiple behaviours were predicted from the same attitudes (e.g. financial contribution to a church, owning a bible, practising Christian beliefs, etc.), the correlations increased to 0.61–0.71, depending on the method used. This has practical implications for studying stockperson behaviour, because it suggests that a variety of stockperson behaviours need to be observed, not merely a single handling bout with an animal, for example.

A major development in the conceptualization of the relationship between attitudes and behaviour came with Ajzen and Fishbein's theory of reasoned action (Ajzen and Fishbein, 1977, 1980). This theory was developed to deal with behaviours that were under the person's control – in other words, volitional behaviours. The theory proposed that the three components of attitude discussed earlier – affect, belief and conation – can better be considered as three response tendencies that represent a sequence in the development of behavioural outcomes. More specifically, the beliefs that people hold, when combined with their evaluations of those beliefs, lead to the formation of attitudes. Intentions and actions then follow from these attitudes. A comprehensive picture of a later revision of the model can be seen in Fig. 4.1 (Ajzen and Fishbein, 1980). From Fig. 4.1, it can be seen that the immediate cause of a person's behaviour is intention. So long as there are no impediments to intention being translated into behaviour, the theory serves as a theory for predicting behaviour. In other words, if there are no physical constraints, such as inability to perform a behaviour or lack of access to the behavioural situation, a person is likely to do what he or she intends.

The immediate cause of intended behaviour is a person's attitude toward the behaviour in combination with the person's subjective norms with respect to the behaviour. A person's subjective norms refer to the extent to which a person believes that relevant other people would approve of the behaviour and the extent to which the person feels willing to comply with other people's expectations. One important feature of this part of the theory is that the object of the attitude is not some general person (or animal) but a behaviour. The theory of reasoned action relies on attitudes towards specific behaviours rather than objects for prediction of acts. This is a major departure from the earlier approaches (for example, the Yale studies by Hovland *et al.*, 1953).

Attitudes are, in turn, determined by a combination of beliefs about the outcomes that are likely to occur following a particular behaviour and an evaluation of those outcomes. For example, if I believed that smoking led to lung disease and I thought that lung disease was a particularly bad outcome, I would have a negative attitude toward cigarette smoking. Similarly, if a child thought that his or her parent would not approve of smoking and the child felt obliged to obey his or her parents, the child would feel a strong subjective normative pressure against smoking.

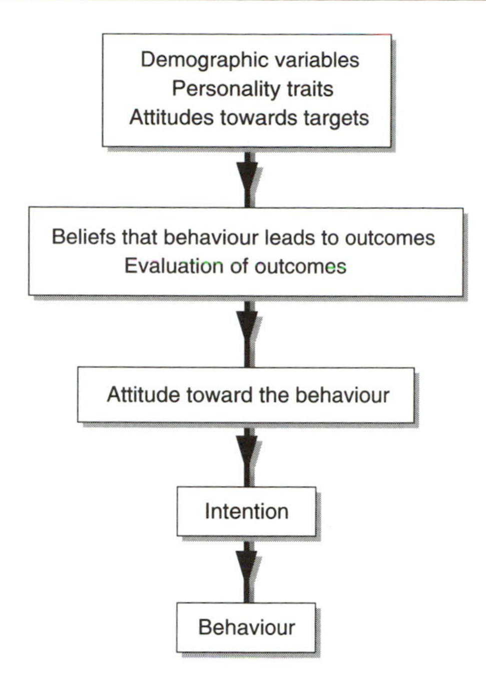

Fig. 4.1. A simplified version of the Ajzen and Fishbein (1980) model of the attitude–behaviour relationship.

The antecedents of beliefs, evaluations and motivations are many and varied. As can be seen in Fig. 4.1, demographic variables, various general attitudes and personality traits indirectly affect behaviour through their influence on beliefs, evaluations and motivations. It is important to recognize that the theory of reasoned action proposes that the important dispositional factor in predicting behaviour is attitude and that other dispositional factors, including personality, operate indirectly through attitudes. These other factors will be considered in the next section.

The prediction of behaviour from attitudes based on the theory of reasoned action can be illustrated with an example from an agricultural industry. Lynne and Rola (1987) carried out a study to investigate the relationship between farmers' attitudes towards soil conservation and their soil-conservation behaviours. Farmers were asked to respond to questions about their beliefs regarding soil conservation. For example, attitude questions used included 'Farmers have a responsibility towards all those now living to use soil resources such as not to cause erosion' and 'Crops can be grown without soil, so erosion is irrelevant.' Because Lynne and Rola were interested in the relevance of economic factors in determining farmers' conservation practices, they also asked questions about farmers' beliefs with respect to economic factors in soil conservation. An example of such a belief statement was 'A farmer must seek to maximize profits no matter what the costs in eroded soil or environmental damage.' Farmers responded to each statement, using a five-point evaluative scale, ranging from strongly agree (5) to strongly disagree (1). Both negative and positive statements towards conservation activity were used. Behaviour was measured in terms of whether or not the farmer engaged in at least one conservation practice. In this example, the authors used farmers' beliefs to measure their attitudes towards conservation behaviour. By using many such belief statements, the authors were relying on consistency theory to justify the measurement of an underlying attitude towards soil conservation with belief statements. As was discussed earlier, if a person holds a variety of beliefs that are consistent with each other, it is possible to infer that person's underlying attitude. Lynne and Rola (1987) ensured that, consistent with Ajzen and Fishbein's (1977) model, attitude questions were directed towards conservation behaviour.

Attitude was found to predict conservation behaviour; for example, variation in attitude accounted for about a third of the variation in conservation behaviour. It was also found that those farmers with higher incomes tended to have poorer attitudes towards conservation behaviour. This study is a clear example of how Ajzen and Fishbein's (1977) model of the attitude–behaviour relationship can be applied in a practical situation. Results from studies in which livestock are intensively handled will be discussed later in this chapter.

There has been substantial research to show that the theory of reasoned action is an excellent predictor of behaviour (Eagly and Chaiken, 1993). However, there have been several variants proposed for the theory. One important additional factor is habit (Bentler and Speckart, 1979). There is ample evidence to show that what people do is partly determined by past experience. This includes studies on voting behaviour (Echabe *et al.*, 1988), seat-belt use (Budd *et al.*, 1984) and blood donation (Bagozzi, 1980). Despite this and other evidence that shows that other factors, such as experience and opportunity, may influence behaviour (Eagly and Chaiken, 1993), the theory of reasoned action has provided a solid basis for predicting behaviour from attitudes.

4.8. ATTITUDES AND PERSONALITY

Underlying characteristics of the person are referred to as personality traits. The idea of a personality trait is that it is a relatively enduring characteristic which exerts a general effect on that person's behaviour and which we cannot observe directly but can infer from the person's behaviour. For example, the concept of introversion/extroversion is widely used in psychology. An extroverted person is outgoing, confident and talkative, while an introverted person is shy, timid and withdrawn. These underlying characteristics result in predictable kinds of behaviours. In social situations, the introvert will avoid being in the limelight, will feel uncomfortable and will express negative feelings, while the extrovert will be just the opposite.

In agriculture, it may be the case that there are personality characteristics – for example, degree of empathy or some temperament factors – which predispose people to be good stockpeople. These characteristics are usually assessed by affective statements by the person. Such statements – for example: 'I feel uncomfortable when I see a distressed animal' – are not easily distinguished from attitude statements, but they actually do differ in that they are self-directed, not directed towards some external person or object. There is some evidence to show that personality may be important in determining what makes a good stockperson. For example, Seabrook (1972a, b) reported that the stockperson's personality was related to behaviour of the cows and milk yield of the herd. He found that high milk yield was associated with herds in which the stockpeople were introverted and confident and where the cows were most willing to enter the milking shed and were less restless in the presence of the stockperson.

It may seem that attitudes and personality traits refer to the same thing. In fact, personality traits differ from attitudes in several ways. First, as previously mentioned, personality traits produce thoughts and behaviours which are not directed towards some external object, but which relate to the individual. Attitudes, on the other hand, are specifically defined in terms of external objects or events. Second, attitudes are evaluative, that is, they refer to things in terms of good/bad, liked/disliked, etc. Personality traits, on the other hand, are descriptive; they describe characteristics of the person, which, while they may have behavioural implications, do not have a prescriptive element. Third, personality traits are somewhat enduring. They are considered to be part of the individual and, although there is some argument about the extent to which they are innate, they remain fairly stable from early childhood to adulthood. For example, Olweus (1979) showed that aggression observed in children as young as 6 months of age significantly correlated with aggression in those same children when they reached adulthood. Attitudes, on the other hand, are thought to develop over time and are a result of the experiences of the person. Fourth, because personality traits are enduring, they do not usually respond to efforts to change them, except, perhaps, under the influence of psychoactive drugs or as a result of brain surgery. Attitudes, on the other hand, are susceptible to change. Indeed, the whole advertising industry, political campaigns and pitches by salespersons and the like are all based on the view that attitudes can be changed.

4.9. STOCKPERSON ATTITUDES AND BEHAVIOUR IN AGRICULTURE

Most of the published research to study the relationship between stockperson attitudes and behaviour has been carried out by Hemsworth and Coleman and their colleagues, in the pig industry (Hemsworth *et al.*, 1989, 1994a; Coleman *et al.*, 1998). Based on Ajzen and Fishbein's (1977, 1980) theory of reasoned action, Hemsworth *et al.* (1989) used an attitude questionnaire to obtain information on the behavioural beliefs of stockpeople about interacting with pigs.

The attitude questionnaire was in two parts. The first half contained a series of belief statements about characteristics of pigs and the second half contained statements about interacting with pigs. The rationale for the development of the questionnaire was that the inclusion of general attitudes towards pigs would allow for the possibility that general attitudes might be related to aspects of the stockperson in areas other than behaviour towards pigs. The relationship between general attitudes towards pigs and other variables will be discussed later in the book. Attitudes towards interacting with pigs were assessed because these were most likely to predict stockperson behaviour towards pigs under the Ajzen and Fishbein (1977, 1980) model. The content of the questionnaires was as follows:

1. *Statements about pigs.* This part of the questionnaire comprised 26 questions. Questions required stockpeople to answer on a five-point scale (1 = disagree strongly to 5 = agree strongly) and the questionnaire included items such as 'Pigs are noisy animals' and 'Pigs are stubborn animals.' High scores were therefore associated with negative attitudes.
2. *Attitudes towards interacting with pigs.* This part of the questionnaire consisted of 46 questions assessing the behavioural beliefs of stockpeople. Questions were answered on a seven-point scale and included items such as 'How much physical effort do you need to use when moving gilts in oestrus?' The response categories ranged from a lot (1) to very little (7). Several variants of each type of question were asked, each variant representing a different target animal. The target animals included 'Gilts in oestrus', 'Non-oestrous gilts', 'First-litter sows in oestrus', 'First-litter sows with piglets', 'First-litter sows after weaning', 'Older sows in oestrus', 'Older sows with piglets' and 'Older sows after weaning'. Two additional attitude questions were included: 'How do you feel about frequent patting and stroking of pigs?' and 'What do you think other farmers feel about patting or stroking of pigs?' These questions were answered on seven-point scales ranging from good (1) to bad (7) and wise (1) to foolish (7). All scores were recorded so that a high score indicated a negative attitude.

The 'attitudes towards interacting with pigs' questionnaire is a questionnaire to assess interrelated behavioural beliefs in a way similar to that used by Lynne and Rola (1987), described above. Consistent responses to these behavioural belief questions reflect the stockperson's underlying attitude towards the particular kinds of interactions with pigs.

The nature of the behaviour of these stockpeople towards pigs during routine mating activities, such as moving pigs for mating, conducting oestrus detection and assisting pigs to copulate, was recorded in this study. Negative or aversive behaviours by stockpeople that were recorded included mild, moderate and forceful hits, slaps and kicks, while the positive behaviours

Table 4.1. Some stockperson attitude–stockperson behaviour correlations in the pig industries.

	Correlations between positive behavioural beliefs and negative stockperson behaviour	
	Petting and behaviour	Effort and behaviour
Hemsworth *et al.* (1989)	–0.61**	–0.47*
Data reanalysed from Hemsworth *et al.* (1994a)	–0.55**	–0.12

Correlation coefficients with * = $P < 0.05$ and ** = $P < 0.01$.
Attitudes assessed on the basis of behavioural beliefs, and the variable used to measure negative stockperson behaviour was the percentage of negative tactile interactions used by the stockperson.

included pats, strokes and the hand of the stockperson resting on the back of the animal. Data on stockperson behaviour were collected and collated on the basis of the number of positive and negative behaviours used by the stockperson per pig handled so that absolute numbers of both positive and negative behaviours used per pig handled and the percentage of negative interactions (i.e. the ratio of negative behaviours to the total number of physical interactions (sum of positive and negative)) used by each stockperson could be studied.

In this study, highly significant correlations were found between stockperson attitudes and stockperson behaviour. For example, a positive attitude towards petting pigs by the stockperson, reflected in positive beliefs about the frequency of patting, stroking and talking to the animals while working with them, showed a large negative correlation with the percentage of negative interactions used by stockpeople (Table 4.1). Similarly, a positive attitude to the use of verbal and physical effort (i.e. beliefs that considerable verbal and physical effort are not generally required to handle pigs) was also associated with a large negative correlation with the percentage of negative interactions. In other words, stockpeople with a positive attitude in respect to these items generally displayed a lower percentage of negative behaviour when interacting with pigs. These correlations therefore indicate that stockpeople with a good attitude towards handling pigs exhibited less negative or aversive behaviours towards pigs. In contrast, a poor attitude by stockpeople was associated with a high percentage of negative behaviours. These results are supported, particularly the petting–behaviour correlation, in a reanalysis of data (Table 4.1) from the study by Hemsworth *et al.* (1994a), in which stockpeople were later trained to improve their attitudes and behaviour towards pigs.

Surprisingly similar attitude–behaviour relationships appear to exist in the dairy industry. In a recent study of human–animal interactions at 29

Table 4.2. Stockperson attitude–stockperson behaviour correlations in the dairy industry.

	Correlations between positive behavioural beliefs and negative stockperson behaviour	
	Petting and behaviour	Effort and behaviour
Hemsworth *et al.* (1995)	–0.47**	–0.36*
P.H. Hemsworth and G.J. Coleman (unpublished data)	–0.54**	–0.28

Correlation coefficients with * = $P < 0.05$ and ** = $P < 0.01$.
Attitudes assessed on the basis of behavioural beliefs, and the variable used to measure negative stockperson behaviour was the percentage of negative tactile interactions used by the stockperson.

commercial dairy farms in Australia (Hemsworth *et al.*, 1995), observations on stockpeople indicate that the attitudes of stockpeople about interacting with animals are also predictive of the behaviour of the stockpeople towards their animals. Beliefs about petting and talking to animals and the amount of physical and verbal effort required to move animals were related to the use of negative interactions by the stockperson (Table 4.2). Positive attitudes to the use of petting and the use of verbal and physical effort to handle animals were negatively correlated with the use of negative tactile interactions, such as slaps, pushes and hits. A recent unpublished study on 37 dairy farms in Australia indicates very similar attitude–behaviour relationships. As shown in Table 4.2, positive attitudes towards petting and effort required to handle cows showed moderate to large negative correlations with the percentage of negative interactions used by stockpeople. Therefore, as in the pig industry, stockpeople were likely to use less negative behaviour when handling their animals if they believed that: (i) petting should be frequently used; and (ii) verbal and physical effort should be infrequently used when interacting with animals. Seabrook (1994) has also provided some illuminating reports of unsolicited stockperson attitude statements which appear to be directed towards behaviour, for example, 'Every day it is the same old routine: feed, move pigs, feed, move pigs. When they won't go where you want them to, it's so easy to lash out with the foot, they are so stubborn' (p. 254). Such statements are consistent with stockpeople behaving in a way determined by their attitudes.

Therefore, observations in the pig and dairy industries indicate that the attitude of stockpeople to handling animals is related to the behaviour of the stockpeople towards their animals. Key beliefs appear to be those relating to petting and use of verbal and physical effort, with a positive attitude to these behaviours associated with less negative tactile behaviour directed towards farm animals. Therefore, these observations indicate that the attitude of the stockperson towards farm animals may be an influential determinant of how the stockperson behaves towards farm animals.

4.10. CONCLUSION

Since the behaviour of stockpeople may influence the behaviour, productivity and welfare of farm animals, it is important to identify the origins of the behaviour of stockpeople. The theory of reasoned action (Ajzen and Fishbein, 1980) holds that:

> as a general rule, we intend to behave in favourable ways with respect to things and people we like and to display unfavourable behaviours towards things and people we dislike. And, barring unforeseen events, we translate our plans into actions.

There is good evidence in the pig and dairy industries that stockperson attitudes do predict stockperson behaviour towards farm animals. Positive attitudes to the use of petting and the use of verbal and physical effort to handle animals were negatively correlated with the use of negative tactile interactions, such as slaps, pushes and hits. Of course, it is likely that factors other than attitudes will contribute to the prediction of behaviour, and these factors will be discussed further in Chapter 6. A key issue, however, is the way in which these attitudes and behaviour develop, and their consequences for the productivity and welfare of farm animals. This is the subject of the next chapter.

Chapter 5

Stockperson Behaviour and Animal Behaviour

5.1. INTRODUCTION

Handling studies, particularly with pigs and poultry, have shown that human–animal interactions in agriculture may have marked consequences for farm animals, particularly in terms of their behaviour, physiology and productivity. For example, as discussed in Chapter 3, growth and reproductive performance of farm animals may be reduced in situations where fear of humans is high. A chronic stress response is implicated in the effects of fear of humans on farm-animal productivity. The implications of human–animal interactions for the productivity of farm animals highlight the need to study the relationships between stockperson behaviour and animal behaviour.

Because attitudes are the main dispositional factor affecting volitional human behaviour, there may be opportunities to manipulate human–animal interactions in order to influence farm-animal productivity and welfare, by improving the attitude and behaviour of stockpeople towards farm animals. Therefore, in Chapter 4, the nature of attitudes and their relationship to behaviour were discussed, both from a theoretical perspective and in terms of some of the available data from agricultural industries. A key issue is how these stockperson attitudes and behaviours affect the behaviour of farm animals. This is the subject of this chapter.

While there are some instinctive behaviours that animals have – for example, fear of strange or novel stimuli or certain fixed mating patterns in response to a hormonal state during oestrus – many behaviours are learned. Farm animals learn to avoid stockpeople if they have had a history of aversive handling. The basic principles of learning apply equally well to animal learning as to human learning. Because of the particular importance of learning for an understanding of how the attitudes and behaviour of stockpeople affect farm animals and how behaviours in farm animals develop, the concepts of learning and reinforcement need to be spelled out, as do the ways in which conditioning, in its various forms, operates.

5.2. KINDS OF LEARNING

Learning is defined as a relatively enduring change in behaviour that occurs as a result of experience. Behaviour refers to observable actions on the part of individuals and we can measure underlying cognitive events by their expression in behaviour. Attitudes fall into this latter category. Not all relatively enduring changes are examples of learning, because ageing or disease may produce such changes. Experience in the particular behaviour or in a closely related behaviour is central to learning.

The most fundamental type of learning (there are other forms of learning recognized: habituation and sensitization, socialization or imprinting, etc.) involves the development of a process where an event, the stimulus, comes to produce a particular behaviour, the response, under conditions where this would not have occurred previously. This process is called conditioning and has two basic forms: classical and instrumental conditioning. The establishment of this connection between stimulus and response in the process of conditioning requires the presence of a reinforcer.

5.2.1. Reinforcement and motivation

Reinforcement is difficult to define and there have been many arguments in the literature about how to arrive at a non-circular definition. Reinforcement is generally defined as anything that follows a response and that increases the likelihood of the response when the same stimulus is presented again. There are two basic kinds of reinforcers: positive and negative. A positive reinforcer is something that is pleasurable or rewarding. Thus food to a hungry individual or water to a thirsty individual would be positively reinforcing. Negative reinforcers are those which arise when some aversive stimulus is removed. It is the release from the aversive stimulus that has the reinforcing effect. Thus removal of an electric shock or cessation of punishment are examples of negative reinforcers. Often it is fairly easy to determine, *a priori*, whether or not something will act as a reinforcer. For example, those things, such as palatable food, which appear to be universally attractive to a person or another animal would be expected to be reinforcers. However, even in this case, if the person has just eaten a large meal, this may not be the case. It is often not possible to determine whether such apparently obvious reinforcers really are reinforcing without observing the development of stimulus–response sequences. What this means is that the strength of a reinforcer often cannot be determined without considering the motivational state of the person or animal. For example, attempts by a boar to mate a gilt or sow are quite aversive for the female if she is not in oestrus, but courtship and copulation are positively reinforcing if she is in oestrus.

The study of motivation is concerned with identifying the causal factors that are responsible for an animal's behaviour, such as feeding, drinking, copulating and fleeing. It has been recognized by many authors that it is generally necessary to consider explanations for observed behaviour in terms of states within the organism (motivations or drives), experiences and learned associations. Indeed, Toates (1980) states that 'the fact that motivation cannot be observed directly and is dependent upon such things as hormone levels and learning, is perhaps no argument against it having a provisional usefulnesss'. For example, if an animal is hungry, its physiological state will generate a motivation or drive that is directed towards food. It is important to realize that motivations directed towards satisfying the basic physiological needs form only one type of motivational state. Motivational states associated with satisfying the basic physiological needs of an animal, such as food and water, have an innate basis and are relatively non-contentious. In contrast, a motivational state associated with consistency (hence the cognitive-consistency theories discussed in the last chapter) may be learned rather than innate and is relatively controversial. Thus, there is a complex interaction between innate characteristics and learned characteristics in determining motivations. The key issue is that there is a large number of causal factors, which, by affecting arousal in the individual, may increase or decrease the intensity of a reinforcing stimulus.

5.2.2. Conditioning processes

Classical conditioning is the most fundamental form of learning and was discovered in the experiments by the famous Russian scientist Ivan Pavlov last century (Pavlov, 1960). Pavlov had observed that his dog began to salivate before actually being given food. He reasoned that some stimulus other than actual eating must trigger the salivation. He then rang a bell each time he gave the dog food. Subsequently, he found that ringing the bell alone was sufficient to induce salivation. Thus the salivation response had become conditioned to the bell (Fig 5.1). To put it more formally, Pavlov paired an unconditioned stimulus (food) with a conditioned stimulus (bell) to eventually produce an unconditioned response (salivation) to the bell alone. This process of conditioning relies on two principles. First, food is unconditioned – that is, the dog has already established the reinforcing properties of food. Second, the conditioned stimulus (bell) has been singled out from all other possibly relevant stimuli and has been presented systematically in conjunction with the food.

Instrumental (operant) conditioning differs from classical conditioning in that there is no explicit pairing of conditioned and unconditioned stimuli. B.F. Skinner (1969) discovered instrumental conditioning and it is best illustrated by describing his experimental procedure. Skinner described an experiment in which a pigeon was placed in a box with a button on the wall. Pressing the button will cause the release of a food pellet. The bird has not learned that pressing the button will release the food, so it will explore the

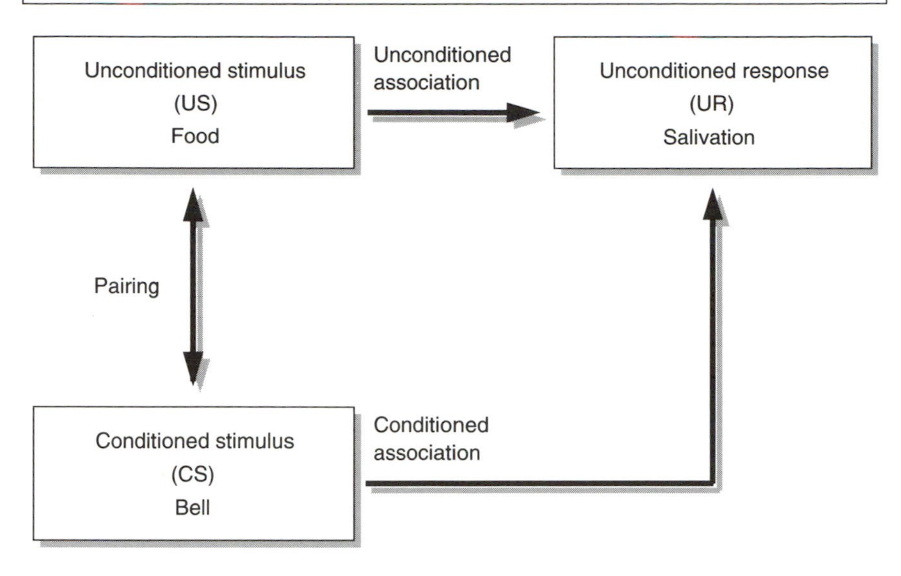

Fig. 5.1. Classical conditioning. Repeated pairing of food and the bell will eventually lead to the bell alone being able to elicit salivation.

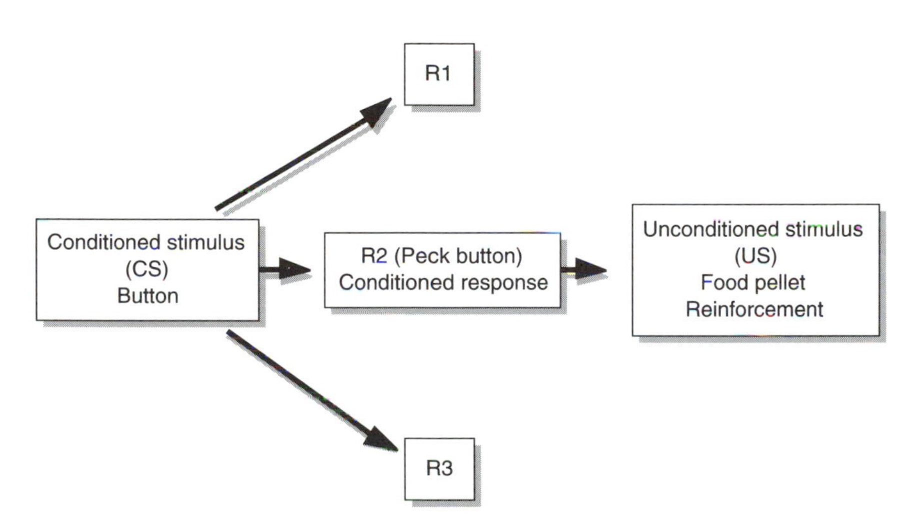

Fig. 5.2. Instrumental (operant) conditioning. When the correct response is elicited, initially by chance, reinforcement is given. Eventually R2 is learned as the correct response.

box. Eventually, by chance, it will peck the button, releasing the food pellet (Fig. 5.2). Because the pigeon has not learned that pressing the button caused the release of food, it will continue with other behaviours until it pecks the button again. Eventually, the contingency between the button and food will become learned and the pigeon will peck the button to get the food. The button will have become operantly conditioned to the food. This form of conditioning differs from classical conditioning in that the individual must discover the stimulus–response contingency.

5.2.3. Primary and secondary reinforcers

When an unconditioned stimulus has pre-existing reinforcing properties, it is called a primary reinforcer. The notion of pre-existing is difficult to define, but food and water are clearly primary reinforcers by this definition. When a stimulus has become conditioned, it may also act as a reinforcer. It is then referred to as a secondary reinforcer because it is not itself a 'natural' reinforcer. In the example of classical conditioning given above, the bell could act as a secondary reinforcer once it had become conditioned. Thus pairing of the bell with, for example, a flash of light could condition the salivation response to the flash of light. In this case, the bell has taken on the reinforcing properties of the primary reinforcer, food, in the original experiment. In fact, most secondary reinforcers do not retain their reinforcing properties indefinitely unless they are regularly paired with the original unconditioned stimulus.

Reinforcement does not need to be constant for learning to take place. In fact, partial reinforcement schedules tend to result in learning that is resistant to later attempts to change the learned response. For example, if, in the classical conditioning experiment described earlier, the food was paired with the bell only 50% of the time, and if the occurrence of the food with the bell was random, the bell would become more strongly conditioned than it would have been in the original experiment.

5.3. RELEVANCE OF LEARNING THEORY TO ANIMAL BEHAVIOUR

5.3.1. Fear of humans

In Chapter 3, we proposed that underlying avoidance behaviour by the animal is the motivational state of fear. The term 'fear' is difficult to define but we shall consider fear as an underlying motivational state that is linked, on the one hand, to particular preceding circumstances or treatments (stimuli) and, on the other hand, to a limited number of types of behaviour (responses).

Level of fear in animals can be inferred from behavioural observations. When an animal is confronted with a fear-provoking stimulus, there is likely to be a number of behavioural responses available to the animal. However, because fear responses function to protect the animal from harmful stimuli, we have proposed that the amount of avoidance of the experimenter or, conversely, the amount of approach to the experimenter is a useful measure of the animal's fear of humans. For example, in studies on pigs, the latency to and the amount of approach to a stationary experimenter in a standard test have been used to measure the level of fear of humans in pigs.

Many animal behaviours are acquired through learning. More specifically, animals learn behaviours, through operant and classical conditioning, which optimize their capacity to live in their environment. For example, there is substantial evidence to show that farm animals learn to avoid stockpeople from whom they have received aversive stimulation (Hemsworth *et al.*, 1989, 1996b; Cransberg and Hemsworth, 1995). In this case, avoiding the aversive handling is acting as a negative reinforcer. Similarly, farm animals show greater approach behaviour towards stockpeople from whom they have received positive interactions (Boissy and Bouissou, 1988; Boivin *et al.*, 1992). Here positive handling is acting as a positive reinforcer.

5.4. STOCKPERSON BEHAVIOUR AND ANIMAL BEHAVIOUR

Over the past 10 years, we have been studying stockperson behaviour–animal behaviour relationships in the animal industries. This research has shown some large and consistent relationships between human behaviour towards animals and the behavioural response of animals to humans across a number of industries, which generally confirm the predictions of numerous handling studies that have been conducted under experimental conditions (see Section 3.5).

Consistent correlations between stockperson behaviour and the level of fear of humans by pigs have been found in the Australian pig industry. The correlations found in two of these studies are presented in Table 5.1. In these two studies, the variable used to measure negative stockperson behaviour was the percentage of negative tactile interactions used by the stockperson, and fear of humans was assessed by the measuring the time spent by pigs near a stationary experimenter in a standard test. Negative tactile interactions by stockpeople include mild to forceful slaps, hits, kicks and pushes, while the positive tactile interactions include pats, strokes and the hand resting on the pig's back. As shown in Table 5.1, the percentage of

Table 5.1. Stockperson behaviour–animal behaviour correlations in the pig industry.

	Negative stockperson behaviour and fear
Hemsworth *et al.* (1989)	0.45*
Coleman *et al.* (1997)	0.40*

Correlation coefficients with * = $P < 0.05$.
Negative stockperson behaviour was assessed by the percentage of negative tactile interactions, while fear of humans was assessed by the time spent near a stationary experimenter.

negative tactile interactions to the total tactile interactions by the stockperson was found to be highly predictive of the level of fear of humans by pigs: high fear levels were observed where stockpeople displayed a high percentage of negative tactile interactions. Surprisingly, high levels of fear of humans were best predicted when the classification of negative behaviours included not only forceful kicks, hits, slaps and pushes, but also negative

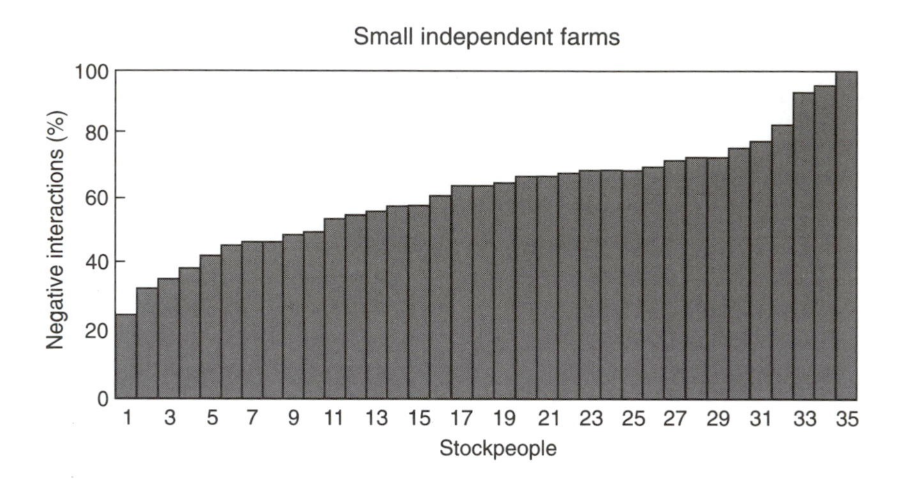

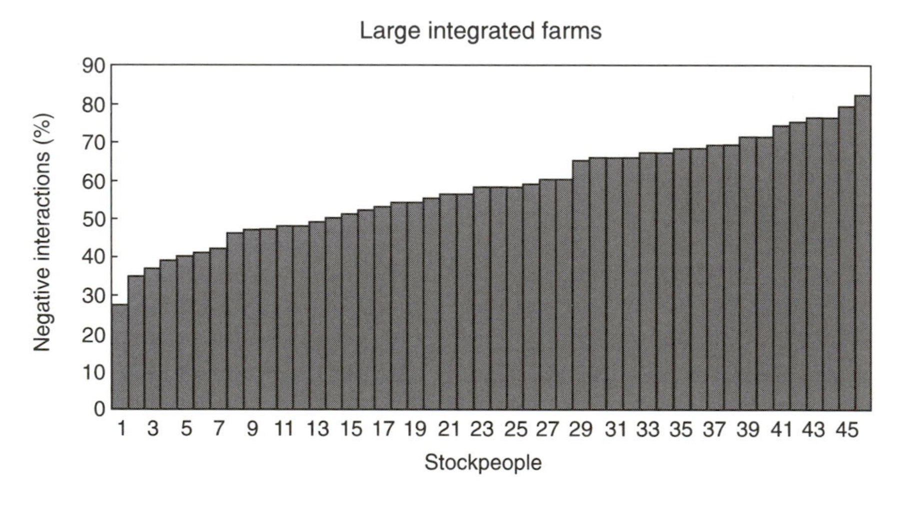

Fig. 5.3. Variation in the behaviour of 35 stockpeople at small independent piggeries and 46 stockpeople at large integrated piggeries (P.H. Hemsworth and G.J. Coleman, unpublished data). The behaviour variable is the percentage of negative tactile interactions directed towards breeding pigs in the mating shed.

behaviours used with less force, such as mild and moderate slaps, prods and pushes. This clearly indicates the sensitivity of pigs to mild and moderate negative interactions by humans, something that is not intuitively obvious to most of us.

The range observed in the behaviour of stockpeople towards breeding pigs in the Australian pig industry is shown in Fig. 5.3. Two types of farms are represented in this figure: small integrated farms, in which one stockperson is predominantly responsible for mating management, and large integrated units, in which teams of stockpeople are responsible for mating management. As can be seen from these data, the range in human behaviour, as assessed by the percentage of negative tactile interactions used moving breeding pigs and when conducting oestrus detection and assisted matings, is very similar at the two types of farms. However, there is a large number of stockpeople at both types of farms that display a high percentage of negative tactile interactions. For example, 57% and 43%, respectively, of stockpeople at small independent and large integrated units consistently displayed more than 60% negative interactions when handling breeding pigs. This high percentage of stockpeople that display predominantly negative interactions towards pigs is of serious concern for the industry and the consequences of this will be considered in detail for all animal industries later in this chapter and in Chapter 6.

It should be appreciated that the pig industry in Australia is typical of intensive pig-production systems in other countries. A comparison of the fear levels observed in Dutch farms with that at Australian farms (Fig. 5.4) indicates the relevance of the Australian data to the international pig industry.

Recent observations in the Australian dairy industry indicate that behavioural patterns of stockpeople, similar to those observed in the pig industry, regulate the fear responses of commercial cows to humans. In two recent studies, involving 29 and 37 farms where cows grazed on pasture all year, moderate to high correlations were found between the percentage of negative tactile interactions used by stockpeople to move cows in and out of the milking shed and the level of fear of humans by cows (Table 5.2). The main negative interactions were pushes, slaps and hits while moving cows into the milking shed and into position for milking, while the main positive interactions were pats, strokes and the hand on the flanks or legs during milking. These associations indicate that fear of humans was high at farms in which stockpeople displayed a high percentage of negative tactile interactions.

The range observed in the behaviour of stockpeople towards dairy cows in the Australian dairy industry is shown in Fig. 5.5. As observed in the pig industry, there is substantial variation in the behaviour of stockpeople towards cows and, because of the relationships between fear and productivity, this variation raises concern for cow productivity and perhaps welfare.

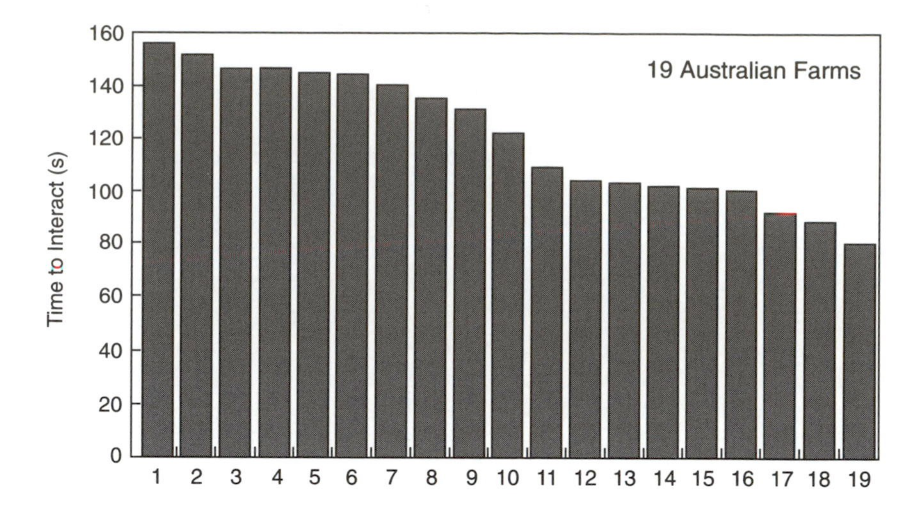

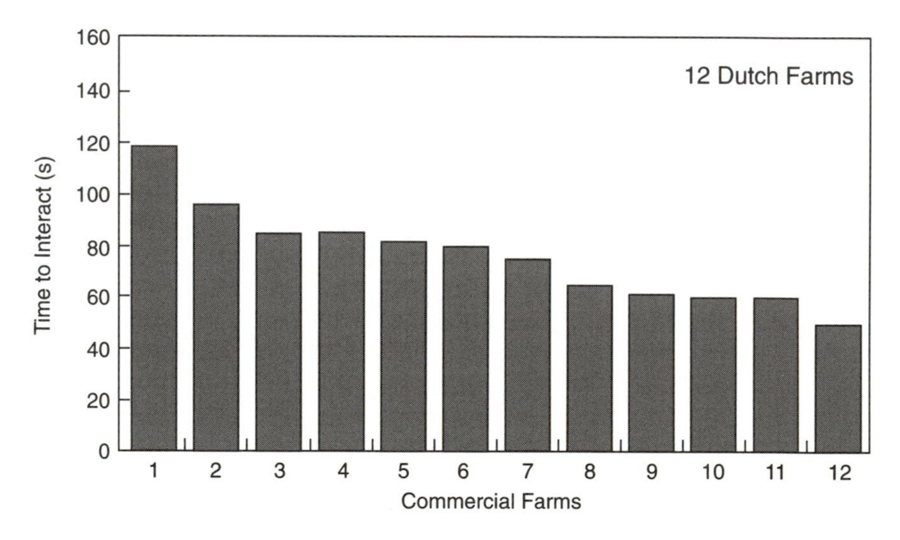

Fig. 5.4. Variation in fear of humans, as assessed by the approach behaviour of pigs to a stationary experimenter at 19 Australian and 12 Dutch piggeries (data reanalysed from Hemsworth *et al.*, 1981b, 1989).

As with experimental studies, these studies with commercial dairy cows and pigs indicate that conditioned approach–avoidance responses develop as a consequence of associations between the stockperson and aversive and rewarding elements of the handling bouts. The main aversive properties of humans include hits, slaps and kicks by the stockperson, while the reward-

Table 5.2. Stockperson behaviour–animal behaviour correlations in the dairy and poultry industries.

	Negative stockperson behaviour and fear
Dairy industry	
Hemsworth *et al.* (1995)	0.31
P.H. Hemsworth and G.J. Coleman (unpublished data)	0.36*
Broiler industry	
Cransberg (1996)	0.43*
Hemsworth *et al.* (1996a)	0.32

Correlation coefficients with * = $P < 0.05$.
Negative stockperson behaviour in the dairy industry was assessed by the percentage of negative tactile interactions, while in the poultry industry it was speed of movement. Fear in cows was assessed by the time spent near a stationary experimenter, while fear of humans in poultry was assessed by the avoidance of an approaching experimenter.

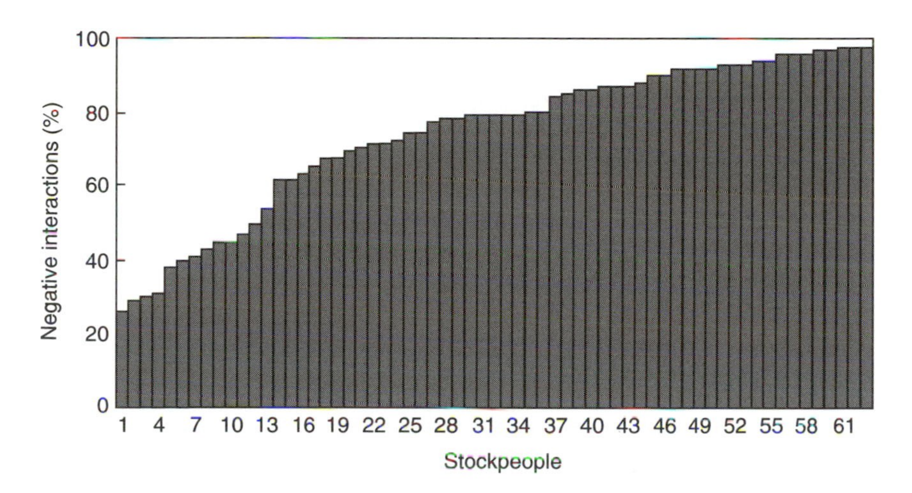

Fig. 5.5. Variation in the behaviour of 63 stockpeople at dairy farms (P.H. Hemsworth and G.J. Coleman, unpublished data). The behaviour variable is the percentage of negative tactile interactions directed towards lactating cows when moving the animals in and out of the milking shed.

ing properties include pats, strokes and the hand of the stockperson resting on the back of the animal. It is the percentage of these negative tactile interactions to the total tactile interactions that appears to determine the commercial animal's fear of humans.

Some limited observations have been conducted recently on human and bird behaviour at 24 and 22 commercial broiler chicken farms. It was found that the behaviour of the stockperson in the shed was moderately associated with levels of fear of humans by birds. As shown in Table 5.2, the speed of movement of the stockperson was positively correlated with the level of fear of humans by birds. Fear of humans was assessed by measuring the avoidance response of birds to an approaching experimenter. These observations indicate that speed of movement by the stockperson is one of the visual behaviours of stockpeople influencing fear of humans at commercial broiler farms. Handling studies also indicate that poultry appear to be particularly sensitive to visual contact with humans and, indeed, positive visual contact may be more effective in reducing levels of fear of humans than human tactile contact (see Section 3.5.2). In addition, sudden and unexpected exposure to the stockperson may be fear-provoking.

Therefore, these observations in the dairy, pig and poultry industries indicate that an important determinant of the fear response of farm animals to humans is the behaviour of the stockperson towards the animals. The large variation in the behaviour of stockpeople towards humans raises some serious concerns for these farm animals and this aspect is briefly considered in Section 5.6.

5.5. FARM ANIMALS' DISCRIMINATION BETWEEN HUMANS

We have proposed in this book that the history of interactions between the stockperson and the animal determines the subsequent stimulus properties of the human for the animal. The characteristics of these stimulus properties include the extent to which the animal has habituated to the presence of the stockperson and the extent to which the stockperson has been associated with rewarding or aversive events. Furthermore, we have previously suggested that, through the process of stimulus generalization, the behavioural response of a farm animal to an individual human will extend to all humans (Hemsworth *et al.*, 1993). Stimulus generalization can be defined as a tendency for stimuli similar to the original stimulus in a learning situation to produce the response originally acquired (Reber, 1988).

Evidence from a number of handling studies supports this view that the animal's response to a single human might extend to include all humans through this process of stimulus generalization. For example, pigs that previously were briefly handled five times per week for 6 weeks by either a handler in a predominantly negative manner or two handlers who differed markedly in the nature of their behaviour towards pigs showed similar behavioural responses to familiar and unfamiliar handlers (Hemsworth *et al.*, 1994b). Similarly, Jones (1994) found that young chickens briefly handled twice daily from 1 day of age in a positive manner by a handler showed similar behavioural responses to the familiar and unfamiliar handlers wearing either similar or different clothing when tested at 10 or 11 days of age.

Studies conducted on commercial pigs and poultry also indicate that farm animals in commercial conditions may not discriminate between familiar and unfamiliar humans. Barnett *et al.* (1993) found that the avoidance responses of commercial laying hens to humans varying in a number of attributes, such as sex, height, clothing and wearing spectacles, were unaffected. The approach behaviour of adult pigs to a familiar handler and an unfamiliar human were similar in a standard approach test conducted at a one-person piggery (Hemsworth *et al.,* 1981b) and Bouissou and Vandenheede (1995) reported that the avoidance responses of sheep towards a human and a human-like model were similar but greater than the response to a novel cylindrical model bearing little resemblance to a human.

Therefore, these results suggest that, in commercial situations, the behavioural response of commercial livestock to one handler may extend to other humans. However, it is possible that there are handling situations in which commercial livestock may not exhibit stimulus generalization. We have previously suggested (Hemsworth *et al.*, 1993) that, in situations in which there is intense handling, predominantly of a positive nature, by a stockperson, animals may learn to discriminate between this handler and other handlers to which the animals may be subsequently exposed. For example, several studies with rodents indicate that animals can discriminate between the caretaker, with whom the animals have had substantial contact, presumably of a positive nature, and a stranger (see review by Dewsbury, 1992). Following an extensive period of intense human contact (10 min day^{-1} for 3 weeks), Tanida *et al.* (1995) found that young pigs showed greater approach to the familiar handler than to an unfamiliar handler, even though both handlers wore similar clothing.

Furthermore, in situations in which the physical characteristics of the handlers may differ markedly, farm animals may learn to discriminate between the handlers. For example, in a series of experiments, de Passille *et al.* (1996) found that dairy calves exhibited clear avoidance of a handler that had previously handled them in a negative manner in comparison with either an unfamiliar handler or a handler that had previously handled them in a positive manner. In order to increase the chances of discrimination, handlers wore different-coloured clothing. Initially there was a generalization of the aversive handling, with calves showing increasing avoidance of all handlers, but with repeated treatment calves discriminated between handlers, and in particular between the 'negative' and 'positive' handlers. It is of interest that discrimination was greatest when tested in the area in which handling had previously occurred rather than in a novel location. In fact, in one experiment, when animals were tested in a location in which the handling was not performed, 40% of the calves actually approached and interacted with the negative handler. These data on calves indicate that discrimination between people by farm animals will be easier if the animals have some distinct cues on which they can discriminate, such as colour of clothing or location of handling.

Although several species of farm animals are capable of discriminating between stockpeople, they do not appear to do so under normal commercial circumstances. Certainly, it is necessary to take into account the number of stockpeople, the environment in which handling occurs, the distinctive features of the stockpeople and the intensity and duration of interactions before being able to predict that farm animals will discriminate between stockpeople. Nevertheless, even when farm animals learn to discriminate between humans, fear responses to humans in general are likely to increase in response to the most aversive handler (Hemsworth *et al.*, 1994b; de Passille *et al.*, 1996). Such a finding has important implications in situations in which several stockpeople may interact with farm animals.

5.6. IMPLICATIONS OF HIGH FEAR OF HUMANS FOR FARM ANIMALS

Some of the implications of high fear of humans for farm animals have been considered in Chapter 3 and will be considered further in Chapter 6; however, it is useful at this stage to briefly review the potential adverse effects of high fear levels.

5.6.1. Fear and productivity

As considered in Section 3.6, there is evidence from both handling studies and observations in the animal industries that high fear levels may reduce the productivity of dairy cows, pigs and poultry. The mechanism(s) responsible is unclear in some species, but studies on pigs show that, in fearful pigs, a chronic stress response or even a series of acute stress responses in the presence of humans may be responsible for the depressed productivity in these fearful animals. Support for this is provided by the known effects of corticosteroids on nitrogen balance, energy balance, key reproductive events and immunosuppression.

5.6.2. Fear and welfare

Fear of humans may also be indicative of reduced welfare of farm animals. The concern for the welfare of fearful animals arises because of injuries that they may sustain in trying to avoid humans during routine inspections and handling, the evidence that, when they regularly interact with humans, these animals are likely to experience a chronic stress response and, finally, the effects of this chronic stress response on immunosuppression, which in turn may have serious consequences for the health of the animals. Additional concerns for the welfare of these fearful animals will be considered in Chapter 6.

5.6.3. Fear and ease of handling

There is evidence in pigs and other farm animals that highly fearful animals are generally the most difficult to handle. As shown in Table 5.3, moderate

Table 5.3. Correlations between the behavioural response of pigs to an experimenter in a standard test to assess fear of humans and the ease of movement of 24 pigs along an unfamiliar route by an unfamiliar handler (Hemsworth *et al.*, 1994b).

	Variables recorded in ease of movement test		
	Time to move	Baulks	Score
Variables recorded in test to assess fear of humans			
Time to approach experimenter	0.34	0.44*	–0.63**
Number of interactions with experimenter	–0.42**	–0.42*	0.51*

Correlation coefficients with * = $P < 0.05$ and ** = $P < 0.01$.
Score was given based on ease of movement, with 0 reflecting substantial difficulty and 4 reflecting little or no difficulty in moving the pig.

to large correlations have been found between the behavioural response of pigs to humans and their ease of handling. These correlations indicate that pigs which showed high levels of fear of humans, based on their avoidance behaviour of an experimenter in a standard test, were the most difficult pigs to move along an unfamiliar route. These fearful pigs took longer to move, displayed more baulks and were subjectively scored as the most difficult to move by the handler. Animals are generally wary of entering an unfamiliar location and, if they are fearful of both the new environment and the handler, they are likely to show exaggerated responses to handling. For example, they may baulk or flee back past the handler. Other studies also indicate that high levels of fear of humans, as a consequence of either reduced human contact or human contact of a negative nature, will decrease the ease of handling of pigs (Gonyou *et al.*, 1986; Grandin *et al.*, 1987). A number of studies on cattle, sheep and horses have shown that handling, generally involving speaking and touching the animals, particularly during infancy, improved their ease of handling (Gonyou, 1993; see also Section 3.5.1).

Observations in the dairy industry (Hemsworth *et al.*, 1995) also indicate that the cows that are the most fearful of humans are the most restless during milking. Significant positive correlations were found between the level of fear of humans by cows and the amount of restlessness, in the form of flinch, step and kick responses in the presence of the stockperson, shown by cows during milking.

5.7. CONCLUSION

As with experimental studies, observations on commercial pigs and dairy cows indicate that conditioned approach–avoidance responses develop as a consequence of associations between the stockperson and aversive and rewarding elements of the handling bouts. The main aversive properties of humans, which will increase the animal's fear of humans, include hits, slaps and kicks by the stockperson, while the rewarding properties, which will decrease the animal's fear of humans, include pats, strokes and the hand of the stockperson resting on the back of the animal. It is the percentage of negative tactile interactions to the total tactile interactions that appears to be the main determinant of the commercial animal's fear of humans. Poultry appear to be particularly sensitive to visual contact with humans, and observations at broiler-chicken farms indicate that rapid speed of movement by the stockperson is fear-provoking for chickens.

The present chapter and Chapters 3 and 4 have considered the influence of human and animal factors on the productivity and welfare of farm animals. The next chapter will integrate these discussions in order to identify opportunities to manipulate human–animal interactions to influence farm-animal productivity and welfare.

Chapter 6

A Model of Stockperson–Animal Interactions and Their Implications for Animals

6.1. INTRODUCTION

The relationships between fear of humans and productivity of farm animals observed in agriculture indicate opportunities to improve animal productivity and welfare. For example, identifying and consequently addressing the factors regulating fear of humans may enable reductions in fear of humans, with possible improvements in the productivity and welfare of farm animals. Handling studies on farm animals under laboratory conditions have shown that human factors are potentially influential in affecting the fear responses of animals to humans. In particular, tactile and visual interactions appear to be particularly important in affecting fear responses. While genetic effects may be influential for naïve animals, experience with humans may ameliorate, maintain or exacerbate the initial fear responses of animals to humans.

Studies on stockpeople in the animal industries have identified some of the influential human characteristics under commercial conditions that affect the behavioural response of farm animals to humans. If fear responses are to be manipulated, it is these human factors and their antecedents that have be targeted for improvement. In Chapters 3, 4 and 5, the influence of attitudes on human behaviour and of stockperson behaviour on farm animal behaviour and productivity were considered and this chapter will integrate the findings and present the key factors involved in the relationships between these human and farm-animal variables, with the objective of presenting a model of human–animal interactions in agriculture. The chapter will form the basis of an introduction to the next two chapters, which deal with the opportunities to change stockperson attitudes and behaviour, utilizing the framework outlined in this model.

6.2. RELATIONSHIPS BETWEEN HUMAN AND ANIMAL VARIABLES

As discussed in previous chapters, there is good evidence, based on handling studies and observations in the animal industries, that human–animal interactions may markedly affect the productivity and welfare of farm animals. This research has provided evidence of a sequential relationship between stockperson attitudes, stockperson behaviour, animal behaviour and animal productivity and welfare, and handling studies have provided evidence that these relationships in the industry may have a causal basis. This section will summarize and extend the previous discussion on these sequential relationships between human factors and key animal variables.

6.2.1. Stockperson attitudes and behaviour

Observations on stockpeople in the pig and dairy industries indicate that the attitudes of stockpeople towards interacting with their animals are predictive of the behaviour of the stockpeople towards their animals. Questionnaires were used to assess attitudes of the stockpeople on the basis of the stockpeople's beliefs about their behaviour and the behaviour of their stock. Positive attitudes to the use of petting and the use of verbal and physical effort to handle animals were negatively correlated with the use of negative tactile interactions, such as slaps, pushes and hits (Table 6.1). These correlations indicate that stockpeople were likely to use a lower percentage of negative tactile behaviour when handling their animals if they believed that: (i) petting should be frequently used; and (ii) verbal and physical effort should be infrequently used when interacting with animals.

These attitude–stockperson-behaviour relationships have not been established in the broiler industry (Table 6.1). Attitudes towards positive physical interactions (petting) with broilers were not recorded in these studies because there is no opportunity for routine tactile interactions between stockpeople and broilers. The correlation between attitude towards the effort required to move birds and stockperson behaviour was small. This is most probably because stockpeople do not routinely move birds as part of the production process. The only exception to this is that, when the birds are very young, stockpeople regularly move them to ensure that the birds are exposed to food and water. It appears that those stockperson interactions with the birds that predict bird behaviour are related to visual cues and include speed of movement. However, attitudes toward these kinds of stockperson behaviour were not assessed. This clearly illustrates the need to identify the stockperson behaviours used as the attitude objects which are relevant to the species being studied. In one of the broiler studies (Cransberg, 1996), positive beliefs about working and interacting with birds were moderately and positively correlated with the time the stockperson spent in the shed.

The generally consistent attitude–behaviour correlations in the pig and dairy industries indicate that the stockperson's attitude towards interacting

Table 6.1. Stockperson attitude–stockperson behaviour correlations in the pig and dairy industries.

	Correlations between positive behavioural beliefs and negative stockperson behaviour	
	Petting and behaviour	Effort and behaviour
Pig industry		
Hemsworth *et al.* (1989)	–0.61**	–0.47*
Hemsworth *et al.* (1994c)	–0.55**	–0.12
Coleman *et al.* (1998)	–0.20	–0.10
Dairy industry		
Hemsworth *et al.* (1995)	–0.47**	–0.36*
P.H. Hemsworth and G.J. Coleman (unpublished data)	–0.54**	–0.28
Poultry industry		
Cransberg (1996)	–	0.03
Hemsworth *et al.* (1996a)	–	–0.13

Significant correlations (* = $P < 0.05$ and ** = $P < 0.01$) indicate associations between the two variables.
Attitudes assessed on the basis of behavioural beliefs, and the variable used to measure negative stockperson behaviour was the percentage of negative tactile interactions used by the dairy and pig stockperson and the speed of movement of the poultry stockperson.

with his/her farm animals may affect his/her behaviour towards these animals. These results demonstrate that one of the antecedents of stockperson behaviour appears to be the attitudes that the stockperson holds towards specific behaviour, and this proposal is underpinned by Ajzen and Fishbein's (1980) theory of reasoned action. As mentioned in Chapter 4, the theory of reasoned action, in brief, is that:

> as a general rule, we intend to behave in favourable ways with respect to things and people we like and to display unfavourable behaviours towards things and people we dislike. And, barring unforeseen events, we translate our plans into actions.
>
> (Ajzen and Fishbein, 1980)

The antecedents of attitudes are many and varied. Demographic variables, various general attitudes and personality traits may indirectly affect behaviour through their influence on attitude towards the behaviour. It is important to recognize that the theory of reasoned action proposes that the important dispositional factor in predicting behaviour is attitude and that other dispositional factors, including personality, operate indirectly through attitudes. The influence of these other factors, either indirectly through attitudes or independently of attitude, on animal productivity is discussed in Section 6.3.

6.2.2. Stockperson behaviour and animal behaviour

As reviewed in Chapter 3, handling studies, predominantly with pigs and cattle, indicate that the level of fear of humans by farm animals is markedly affected by tactile interactions from stockpeople. In these handling studies, fear of humans was generally assessed on the basis of the approach behaviour of the animal to a stationary experimenter in a standard test.

Handling studies on pigs have shown that negative tactile interactions, imposed briefly but regularly by humans, will produce high levels of fear of humans. For example, negative interactions imposed daily for as little as 15–30 s consistently resulted in pigs showing marked avoidance of humans when subsequently tested with a stationary experimenter (see Section 3.5.1). In contrast, brief positive handling resulted in low fear levels. Further evidence of the sensitivity of pigs to human behaviour are the results of the study by Gonyou *et al.* (1986). A regular handling treatment involving no tactile interactions but rapid and close approach by the experimenter resulted in pigs showing marked avoidance of humans, similar to those responses shown by pigs that received a shock from a battery-operated prodder whenever they failed to avoid the approaching experimenter.

Table 6.2. Stockperson behaviour–animal behaviour correlations in the dairy, pig and poultry industries.

	Negative stockperson behaviour and fear
Pig industry	
Hemsworth *et al.* (1989)	0.45*
Hemsworth *et al.* (1994c)	0.01
Coleman *et al.* (1998)	0.40*
Dairy industry	
Hemsworth *et al.* (1995)	0.31
P.H. Hemsworth and G.J. Coleman (unpublished data)	0.36*
Broiler industry	
Cransberg (1996)	0.43*
Hemsworth *et al.* (1996a)	0.32

Significant correlations (* = $P < 0.05$ and ** = $P < 0.01$) indicate associations between the two variables.

Negative stockperson behaviour in the pig and dairy industries was assessed by the percentage of negative tactile interactions, while in the poultry industry it was speed of movement. Fear in pigs and cows was assessed by the time spent near a stationary experimenter, while fear of humans in poultry was assessed by the avoidance of an approaching experimenter.

Observations in the pig industry indicate that the nature of human tactile interactions is an influential factor affecting the behavioural response of commercial pigs to humans. The percentage of negative tactile interactions to the total tactile interactions by the stockperson was positively correlated with the level of fear of humans, as assessed by the amount of approach by pigs to a stationary experimenter (Table 6.2). Negative tactile interactions include moderate to forceful slaps, hits, kicks and pushes, while the positive tactile interactions include pats, strokes and the hand resting on the pig's back. Thus pigs were most fearful of humans at farms in which stockpeople used a high percentage of negative tactile interactions in handling pigs (Fig. 6.1).

As in the pig industry, negative tactile interactions by stockpeople may regulate the fear responses of commercial cows to humans. The percentage of negative tactile interactions to the total tactile interactions used by the stockperson was positively correlated with the level of fear of humans, as assessed by the amount of approach by cows to a stationary experimenter in a standard test (Table 6.2). The main negative interactions were pushes, slaps and hits while moving cows into the milking shed and into position for milking, while the main positive interactions were pats, strokes and hand on the flanks or legs during milking. Thus cows were most fearful of humans at farms in which stockpeople used a high percentage of negative tactile interactions in handling their cows. Interestingly, speed of movement by the

Fig. 6.1. A high percentage of negative interactions used by the stockperson, such as kicking or slapping, will increase the commercial pig's fear of humans.

stockperson and the distance between the stockperson and the last animal in the herd when moving cows to the milking shed from pasture were also predictive of fear levels: both fast speed of movement and a greater distance between the stockperson and the herd were associated with higher levels of fear of humans (Hemsworth *et al.*, 1995).

Experiments by Rushen and colleagues and experiments in our laboratory have also shown that cows will quickly learn to avoid humans using negative interactions. Brief exposure to a handling treatment involving slapping or briefly shocking with a battery-operated prodder by experimenters resulted in cows rapidly showing avoidance responses to humans (Muunksgaard *et al.*, 1995; de Passille *et al.*, 1996). Moderate or forceful slaps imposed briefly whenever heifers failed to avoid the approach of a handler subsequently resulted in reductions in the approach of the heifers to a stationary experimenter in a standard test and an increase in the flight distance of the heifers from an approaching experimenter (Breuer *et al.*, 1997). Regular interactions with feedlot cattle, in which humans slowly approached and squatted to encourage approach by the animals, resulted, presumably through habituation, in reductions in the animals' fear of humans (Hemsworth *et al.*, 1996c). Similarly, a number of handling studies on cattle involving the imposition of positive tactile interactions, such as pats, strokes and fondling, have shown that handled animals displayed less avoidance of humans in a range of testing situations designed to assess the behavioural response to humans (Boissy and Bouissou, 1988; Boivin *et al.*, 1992).

Studies with poultry have shown that chickens and laying hens are particularly sensitive to visual contact with humans. Regular treatments involving the experimenter placing his/her hand either on or in the chicken's cage and allowing birds to observe other birds being handled have been shown to result in reductions in the subsequent avoidance of humans shown by young chickens (Jones, 1993). Interestingly, visual contact without tactile contact was more effective in reducing fear than picking up and stroking the bird, suggesting that this tactile handling by humans may contain aversive elements for birds, such as active interaction and tactile interaction. A handling study on laying hens by Barnett *et al.* (1994) also clearly demonstrates the influential effects of visual contact with humans on fear responses of birds to humans. Regular visual contact with humans, involving positive elements, such as slow and deliberate movements by the experimenter, markedly reduced the subsequent avoidance behaviour of mature laying hens to humans in comparison with minimal human contact that at times contained elements of sudden and unexpected human contact.

Observations on stockpeople at broiler-chicken farms reveal that the visual cues from the stockperson may regulate the fear responses of commercial birds to humans. The speed of movement of the stockperson was correlated with the level of fear of humans by birds (Table 6.2). Frequency of tapping by the stockperson on objects while moving through the shed was positively associated with fear levels, but, surprisingly, frequency of waving by the stockperson to move birds in order to inspect them as the stockperson moved through the shed was positively, not negatively, associated with avoidance by birds of the experimenter. These correlations indicate that birds are most fearful of humans at farms in which stockpeople move quickly through the shed, frequently tap on objects in the shed and infrequently wave as they move through the shed. It is possible that waving by the stockperson, which intuitively appears to be fear-provoking, may be either rewarding, mildly fear-provoking, resulting in rapid habituation of the fear responses to humans, or a consequence of birds remaining closer to the stockperson as he/she moves slowly through the shed, necessitating the frequent use of this behaviour to move birds from under the stockperson's feet. Furthermore, this distinct pattern may alert birds to the imminent approach of the stockperson and reduce the chances of unexpected exposure, which in itself may be highly fear-provoking. In contrast, speed of movement appears to be highly fear-provoking and the tactile contact that accompanies high speed of movement by the stockperson and the corresponding avoidance responses (flight and vocalization) by birds that receive tactile contact from the stockperson may exacerbate these fear responses throughout the flock.

These studies with farm animals indicate that conditioned approach–avoidance responses develop as a consequence of associations between the

Fig. 6.2. Positive interactions by the stockperson, such as the hand resting on the cow's back, will reduce the commercial cow's fear of humans.

stockperson and aversive and rewarding elements of the handling bouts. For pigs and cattle, the main aversive properties of humans, which will increase the animal's fear of humans, include hits, slaps and kicks by the stockperson, while the rewarding properties, which will decrease the animal's fear of humans, include pats, strokes and the hand of the stockperson resting on the animal (Fig. 6.2). For poultry, the main aversive properties of humans, which will increase the animal's fear of humans, appear to include fast speed of movement and unexpected movement or appearance of the stockperson. As mentioned in Chapter 3, little is known of the effects of other forms of human contact.

6.2.3. Animal behaviour and animal productivity and welfare

Handling treatments inducing high levels of fear of humans have been shown to markedly reduce the growth and reproductive performance of pigs (see Section 3.6.1). Furthermore, observations in the pig industry have revealed negative correlations, based on farm averages, between fear of humans and reproductive performance of pigs (Table 6.3). These correlations indicate that the productivity of pigs was lowest at farms in which pigs were most fearful of humans. It is noteworthy that Hemsworth *et al.* (1989) found that variation in fear of humans accounted for about 20% of the variation in

Table 6.3. Animal behaviour–animal productivity correlations in the dairy, pig and poultry industries.

	Fear and animal productivity
Pig industry	
Hemsworth *et al.* (1981b)	–0.51*
Hemsworth *et al.* (1989)	–0.55*
Hemsworth *et al.* (1994c)	–0.01
Dairy industry	
Hemsworth *et al.* (1995)	–0.46*
Broiler industry	
Hemsworth *et al.* (1994a)	–0.57**
Cransberg (1996)	–0.10
Hemsworth *et al.* (1996a)	–0.39
Egg industry	
Barnett *et al.* (1992)	–0.58**

Significant correlations (* = $P < 0.05$ and ** = $P < 0.01$) indicate associations between the two variables.
Fear of humans by pigs and cows was assessed by the time spent near a stationary experimenter, while fear of humans by poultry was assessed by the avoidance of an approaching experimenter. The productivity variables were reproduction in pigs, milk yield in cows, feed conversion in chickens and egg production in hens.

reproductive performance across the study farms, indicating the importance of fear of humans with regard to pig productivity.

The mechanism responsible for these adverse effects of high fear on productivity appears to be a chronic stress response, because, in a number of experiments on pigs, handling treatments that resulted in high fear levels also produced either a sustained elevation in the basal concentrations of the stress hormone cortisol or enlargement of the adrenal glands, together with depressions in growth and reproductive performance (see Section 3.6.1).

Handling of a negative nature has generally been shown to reduce the growth performance of chickens. Gross and Siegel (1979, 1980, 1982) found that young chickens that received frequent human contact, apparently of a positive nature, had improved growth rates and feed efficiency and were more resistant to infection than birds that received minimal human contact or birds that had been deliberately scared. Deliberate scaring in the third treatment involved shouting and banging on the birds' cages. In a study with adult laying hens, Barnett *et al.* (1994) found that regular visual contact, involving positive elements, such as slow and deliberate movements, which reduced the subsequent avoidance behaviour of mature laying hens, resulted in higher egg production than a treatment which involved minimal and at times negative human contact. These results and those of other studies provide support for the proposition that handling treatments which increase fear of humans will depress the productivity of poultry (see Section 3.6.2).

Field observations on poultry have found significant negative relationships, based on farm averages, between the level of fear of humans and the productivity of commercial broiler chickens and laying hens (Table 6.3). Both egg production of laying hens and the efficiency of feed conversion of broiler chickens were negatively correlated with the level of fear of humans by birds; high fear levels were associated with low productivity. Bredbacka (1988) also reported that egg-mass production was lower in hens that showed increased avoidance of the human.

Recent observations on commercial dairy cows also indicate the existence of a significant negative fear–productivity relationship (Table 6.3). The approach behaviour of dairy cows to an experimenter in a standard test was positively correlated with milk yield of the farm, indicating that milk yield was lowest at farms in which cows were highly fearful of humans. Seabrook (1972a) has also suggested that milk yield may be at risk when cows are fearful of humans.

Therefore, handling studies and observations in the animal industries on fear–productivity relationships indicate that high levels of fear of humans may limit the productivity and welfare of farm animals. One possible mechanism responsible for these effects is a chronic stress response, since, in a number of handling studies, a sustained elevation in corticosteroids was found in animals showing high fear levels.

6.3. MODEL OF HUMAN–ANIMAL INTERACTIONS IN AGRICULTURE

As a consequence of this research on human–farm-animal interactions, the following model of human–animal interactions in agriculture has been proposed (Hemsworth *et al.*, 1993). Because a stockperson's behaviour towards animals is largely under his/her control, this behaviour is strongly influenced by the attitudes that the stockperson holds about the animals. These attitudes and consequent behaviours predominantly affect the animal's fear of humans, which, in turn, affects the animal's performance and welfare. The mechanism whereby fear affects performance and welfare appears to be through a chronic stress response. These sequential relationships between human and animal variables are depicted in Fig. 6.3.

Based on handling studies, there is evidence to implicate a causal basis for the relationships between stockperson behaviour, animal fear and animal productivity, and this aspect will be considered further in the next chapter. This sequential relationship is not necessarily a surprise to some, particularly those working in agriculture, but the magnitude and consistency of the relationships across a number of industries are surprising. The fact that studies in the dairy, pig and poultry industries indicate that fear of humans may account for a fifth to a third of the variation in productivity seen across farms in these industries demonstrates the importance of human–animal interactions for farm-animal productivity.

In our earlier discussion of the relationship between attitudes and behaviour, we mainly considered the role of attitudes as dispositional factors in determining behaviour where such behaviour is under the volitional control of the person. However, as we shall see in Chapter 7, the behavioural situation feeds back on attitudes. So, for example, the mere fact that a stockperson behaves in a particular way will tend to reinforce that stockperson's attitudes towards that behaviour. In addition, the outcome of the behaviour will also feed back on the stockperson's attitudes. Thus, if a stockperson has a negative attitude towards handling pigs and behaves negatively toward

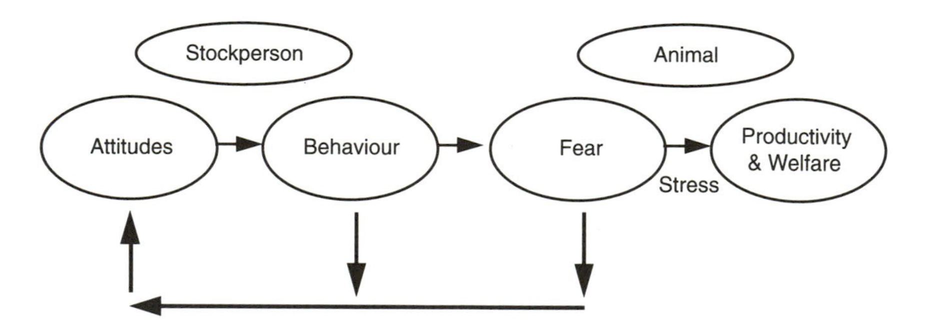

Fig. 6.3. A model of human–animal relationships in animal production.

pigs, thus making them more difficult to handle (because of increased baulking, fleeing, slipping, etc.), this will reinforce the stockperson's original attitudes.

This feedback loop in the model provides an opportunity to modify stockperson attitudes and behaviour. In Chapter 7, we shall discuss in detail how this feedback loop can be used to train stockpersons by targeting these key attitudes and behaviour simultaneously.

The major task of stockpeople in modern animal-production systems is to care for farm animals by managing their social, climatic, nutritional and health requirements to ensure their optimal productivity and welfare. Such a responsibility requires stockpeople to have appropriate technical knowledge and skills to provide the animal in a timely and efficient manner with the correct conditions to survive, grow and reproduce at a high level. Therefore, it is obvious that the stockperson's knowledge and skills in animal production, together with the motivation to utilize these attributes, i.e. work motivation, are important determinants of the productivity and welfare of his/her stock. Managers and employers of stockpeople can influence the knowledge and skills of their stockpeople by encouraging and providing opportunities for their staff to participate in external and internal training programmes, while also providing the working conditions and rewards to encourage their staff to utilize their knowledge and skills. Job satisfaction is widely regarded as an important factor affecting work motivation. This pathway, in which a stockperson's technical knowledge and skills may influence animal productivity and welfare, can be addressed by the provision of appropriate training courses for stockpeople. Developments in defining the necessary knowledge and skills required in each animal industry and aligning remuneration and promotional opportunities with these aspects will not only promote technical training for stockpeople but provide a range of benefits, such a recognized career and remuneration pathway, while promoting self-esteem in stockpeople and the professionalism and image of the industries.

In trying to integrate the stockperson attitudes and behaviour with these other job-related characteristics, it is not too difficult to recognize that the attitude of the stockperson to the subject of his work, the animal, may influence a number of these other influential characteristics and thus the work performance of the stockperson. For instance, the attitude of the stockperson towards the animal may affect such job-related characteristics as work motivation, motivation to learn new skills and knowledge about the animal and job satisfaction, which in turn may affect work performance of the stockperson. In many industries outside agriculture, motivating factors which appear to be important determinants of job satisfaction and thus in turn work motivation have traditionally been considered to include achievement, recognition, responsibility, the work itself and advancement. Other factors, often called hygiene factors, which may not actually increase job satisfaction but, when they are at suboptimal levels, act to depress job satisfaction, include company policy, pay, working conditions and benefits. Therefore, clearly, if the stockperson's attitude towards interacting with pigs, the subject of the stockperson's tasks, is poor and, in conjunction with the consequent effects that this may create, such as handling, production and welfare difficulties, the stockperson's job satisfaction is likely to deteriorate, with adverse consequences for work motivation. Furthermore, if the stockperson's attitude towards the animal is poor, the stockperson's commitment to the surveillance of and the attendance to production and welfare problems facing the animal is likely to deteriorate. Thus, the attitudinal and behavioural profiles of the stockperson may have marked effects on animal productivity and welfare, both via fear of humans by the animal and via work performance of the stockperson.

In fact, some recent research in the Australian pig industry (Coleman *et al.*, 1998) has indicated relationships between the stockperson's attitudes and a number of job-related variables. In this study, four scores were obtained from questions relating to aspects of the job. The first score, labelled 'Job enjoyment', comprised five items, including 'How boring is your job?' and 'How long do you think you will continue in the pig industry?' The second score, labelled 'Family', came from two items: 'When you go away for holidays, do you go with a member of your family?' and 'Who normally supervises your children when you are not able to?' The third score, labelled 'Work breaks', was based on two items: 'How much do you look forward to tea and lunch breaks?' and 'How much do you look forward to the end of the working day?' The fourth score, labelled 'Working conditions', was also based on two items: 'I often have to work in cramped conditions' and 'The air is clean at work.' High scores meant high job satisfaction in all cases.

Two scores were also extracted from those questions relating to technical knowledge and willingness to learn. The first score, labelled 'Learning', came from four questions, including 'How often do you discuss work methods during tea and lunch breaks?' and 'Would you attend training courses in your own time if they were available?' The second score, labelled 'Knowledge', comprised three questions, including 'How much do you know about diseases in pigs?' and 'How much do you know about factors which affect reproduction in pigs?' Low scores meant low technical knowledge and willingness to learn.

It was found that the willingness of stockpeople to attend training sessions in their own time (score labelled 'Learning') was correlated with attitudes towards characteristics of pigs and towards most aspects of working with pigs (Table 6.4).

Job enjoyment and opinions about working conditions showed similar relationships with attitudes (Table 6.4). Thus, the stockperson's attitudes may, indeed, be related to aspects of work apart from handling of animals. The possible interrelationships between attitude towards animals and these other job-related characteristics are depicted in Fig. 6.4.

Table 6.4. Correlations among attitude subscales and work-related variables in the pig industry.

	Attitude subscales						
	Negative belief about pigs	Negative behaviour	Working with pigs	Characteristics of pigs	Pigs as pets	Handling non-oestrous pigs	Handling oestrous pigs
Job enjoyment	–0.30*	–0.24	–0.42**	–0.46**	–0.25	–0.20	–0.23
Work breaks	–0.42**	–0.32*	–0.10	–0.00	–0.15	–0.09	–0.21
Working conditions	–0.20	0.10	–0.32*	–0.30*	–0.03	–0.34*	–0.25
Learning	–0.31*	–0.28*	–0.39*	–0.26*	–0.32*	–0.25*	–0.16
Knowledge	–0.13	–0.02	–0.25	–0.12	–0.38*	0.02	–1.01

Significant correlations (* = $P < 0.05$ and ** = $P < 0.01$) indicate associations between the two variables.
High scores for the attitude subscales indicate a negative belief, while high scores for the work-related variables indicate a positive response.

Table 6.5. Correlations among attitude subscales and empathy in the pig industry.

	Attitude subscales						
	Negative belief about pigs	Negative behaviour	Working with pigs	Characteristics of pigs	Pigs as pets	Handling non-oestrous pigs	Handling oestrous pigs
Feelings	–0.26	–0.27	–0.18	–0.48**	–0.32*	–0.12	0.01
Reactivity	0.08	–0.31*	0.03	–0.42**	–0.20	–0.35*	–0.30*

Significant correlations (* = $P < 0.05$ and ** = $P < 0.01$) indicate associations between the two variables.
High scores for the attitude subscales indicate a negative belief, while high scores for the empathy variables indicate a positive response.

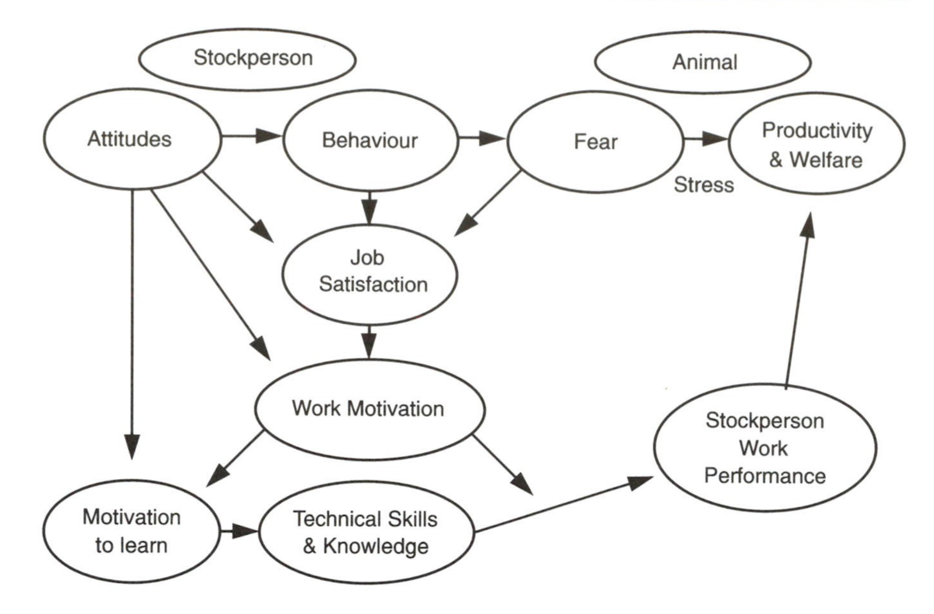

Fig. 6.4. The interrelationships between stockperson attitudes and behaviour and other job-related characteristics.

Several authors have proposed that other stockperson characteristics may influence the performance of stockpeople in agriculture. In a study of 87 stockpeople at a large piggery, Coleman *et al.* (1998) found some modest relationships between empathy and stockperson attitudes, suggesting that empathy may influence the development of beliefs about pigs themselves and about handling pigs (Table 6.5). The 'Feelings' measure consisted of three questions including 'How do you feel when pigs are injured?' and 'To what extent do you think weaners are cute?' The 'Reactivity' subscale comprised three questions, including 'How much do pigs react to you?' and 'I get very upset when I see an animal in pain.' As shown in Table 6.5, there was a general pattern of negative correlations between empathy and negative attitudes towards pigs.

Seabrook (1972a, b) reported that the stockperson's personality was related to both the behaviour of the cows and the milk yield of the herd. In 28 one-person herds, he found that the highest-yielding herds were those where the stockpeople were introverted and confident and where the cows were most willing to enter the milking parlour and were less restless in the presence of the stockperson. Significant relationships have been found in the pig industry between personality types of stockpeople, based on an evaluation using the Sixteen Personality Factor Questionnaire, and productivity in farrowing units of both small independent and larger integrated units (Ravel *et al.*, 1996). At large integrated units, high preweaning performance (number of piglets weaned) was associated with stockpeople with traits of warmth, emotional stability and self-discipline, while low preweaning performance was associated with stockpeople being highly self-assured and sensitive. However, the strongest predictors of the stockperson's behaviour have been found to be the stockperson's attitudes (see Table 6.1). Variables such as confidence, introversion and empathy may modulate the manner in which a stockperson's beliefs, behaviour and their consequences are established (Ajzen and Fishbein, 1980). For example, an insensitive, rigid and overconfident view may limit the extent to which the stockperson learns appropriately from his or her experiences with animal-handling problems, and therefore this view may retard the development and maintenance of appropriate stockperson's beliefs about how animals should be handled. Nevertheless, it has yet to be determined whether such variables as personality and empathy could independently contribute to fear, welfare and productivity in farm animals or could act by modulating attitudes and beliefs, as Ajzen and Fishbein (1980) have proposed. It is of interest that Beveridge (1996), in a study of 44 stockpeople at a large integrated piggery, found no consistent correlations between personality type and attitudes of stockpeople towards characteristics of pigs or towards most aspects of working with pigs. Personality type was assessed by the Myer–Briggs Type Indicator questionnaire in which indices of extroversion–introversion, sensing perception–intuitive perception, thinking judgement–feeling judgement and judgement–perception were scored. This further serves to illustrate that research on stockpeople using a 'big five' measure of personality needs to be done. The Myer–Briggs Type Indicator is designed primarily as a classification tool and is not really appropriate for providing quantitative data on all of the dimensions of personality for the purposes of predicting worker behaviour.

6.4. CONCLUSIONS

Research on the interactions between humans and farm animals has shown interrelationships between the stockperson's attitudes and behaviour and the behaviour, productivity and welfare of farm animals and the following model of human–animal interactions in agriculture has been proposed. Because a stockperson's behaviour towards animals is largely under volitional control, this behaviour is strongly influenced by the attitudes that the stockperson holds about the animals. These attitudes and consequent behaviours predominantly affect the animal's fear of humans, which, in turn, affects the animal's performance and welfare. The mechanism whereby fear affects performance and welfare appears to be through a chronic stress response. The stockperson's attitude and behaviour may have either direct or indirect affects on other important job-related characteristics, such as job satisfaction, work motivation and motivation to learn, and thus also affect animal productivity and welfare. A less obvious risk to the welfare of farm animals arises in situations in which the attitude and behaviour of the stockperson towards the animals are negative because the stockperson's commitment to the surveillance of and the attendance to welfare issues is most probably highly questionable.

The sequential relationships between human and animal variables indicate the opportunity to target stockperson attitudes and behaviour in order to improve animal productivity and welfare. Stockperson selection and training programmes addressing these key attitudinal and behavioural profiles appear to offer the animal industries potential to improve animal productivity and welfare; this is the subject of the next two chapters.

Chapter 7

Changing Stockperson Attitudes and Behaviour

7.1. INTRODUCTION

In the previous chapter, the research on the interactions between humans and farm animals was reviewed and a model of human–animal interactions in agriculture was proposed. Based on this research, it is proposed that sequential relationships exist between the stockperson's attitudes and behaviour and the behaviour, productivity and welfare of farm animals. There are also reciprocal relationships operating within this pathway; it is not unidirectional but bidirectional with animal behaviour – for example, feeding back on the attitudes and thus behaviour of the stockperson. In addition to this pathway affecting animal performance and welfare, it is suggested that the stockperson's attitudes may influence other important job-related characteristics of the stockperson, such as job satisfaction and job motivation, and thus influence animal performance and welfare via this pathway. Studies on both experimental and commercial animals, particularly dairy cattle, pigs and poultry, support this model. Thus these sequential relationships between human and animal variables indicate the opportunity to target stockperson attitudes and behaviour in order to improve animal productivity and welfare.

The aim in this chapter is to review the principles underlying attitude and behavioural change and to discuss the research in agriculture in which attitude and behavioural change has been attempted.

7.2. FACTORS INFLUENCING STOCKPERSON ATTITUDES AND BEHAVIOUR

In order to understand how the attitudes and behaviour of stockpeople can be changed, it is necessary to appreciate the factors that influence the establishment and maintenance of attitudes and behaviour.

7.2.1. Learning and attitude acquisition

Stockpeople, through both classical and instrumental conditioning and in the context of varying motivational states, acquire their attitudes and behaviour towards the animals under their care. The processes of instrumental and classical conditioning in animals, which were discussed in Chapter 5, also apply to human learning, and thus experiences, particularly those when first interacting with animals or first observing others interacting with animals, are influential in the acquisition of attitudes and behaviour towards farm animals. However, humans can also learn indirectly by observing behaviour in others rather than by direct conditioning processes. Such processes are particularly relevant to understanding the way in which attitudes and behaviours develop in stockpeople.

Kanekar (1976) proposed that attitudes can be acquired by conditioning without any firsthand experience of the attitude object. This rests on the assumption that people can vicariously experience the emotions of others. Because these emotional experiences can act as reinforcers, mere observation of an attitude object and another person's emotional response to that object is sufficient for the observer to develop an attitude by classical or operant conditioning, as though the observer was directly involved in the experience. An example of this is as follows. Suppose that a person observes another person expressing a strong negative attitude towards a particular farm animal, say a pig, and at the same time shows clear disgust towards pigs. By classical conditioning, the observer would experience a pairing of the object 'pig' with the vicariously experienced emotion of disgust. If this was repeated sufficiently often, the observer would develop a negative attitude towards pigs.

The direct evidence in support of Kanekar's view is scarce. Nevertheless, there is widespread evidence that people can and do learn attitudes through observation. Maio *et al.* (1994) carried out a study in which people were given information about a new immigrant group. Two variables were systematically varied. First, the immigrants were described in a way that made them either very relevant to the participants in the study, or largely irrelevant. Second, participants were told about others' emotional reactions to the immigrants. In two of the experimental conditions, one group was given descriptions of positive reactions and the other group was given descriptions of negative reactions. When consistently positive emotional descriptions were provided and personal relevance was high, participants formed more positive attitudes than when consistently negative emotions were provided.

It is clear, from the foregoing, that people can acquire attitudes indirectly without having any direct experience with the attitude object. Of course, direct experience with an attitude object can also lead to the development of attitudes through learning.

7.2.2. Effects of behaviour on attitudes

One of the major contributions of the theory of cognitive dissonance (Festinger, 1957), briefly described in Chapter 4 (Section 4.5), was its emphasis on the reciprocal relationship between attitudes and behaviour. This refers to the fact that not only do attitudes influence behaviour, but also behaviour influences attitudes. Once a person carries out a particular behaviour, there is a tendency to modify those attitudes which are relevant to performing that behaviour. The classic study in this area was reported by Festinger and Carlsmith in 1959. They asked undergraduate students to spend 2 h performing a very boring task. The students were then paid either nothing, $1.00 or $20.00 (a lot of money in 1957, when the experiment was done). Prior to being paid, each student who was to be paid was asked (persuaded really) to tell another person that the experiment had been very interesting and that the person would enjoy it. All students were subsequently interviewed and asked whether they enjoyed the experiment.

So, basically, students were told to tell another person that performing a very boring task was, in fact, not boring at all. In other words, those students who received either $1.00 or $20.00 were asked to lie about how enjoyable the task was. Those who were paid nothing were not told anything about the task. Curiously, those who were paid $1.00 said that the experiment was enjoyable and that they would participate in a similar experiment in the future. Those paid nothing or $20.00 said that they did not enjoy the experiment and were less willing to participate in the future. In other words, those students paid nothing told the truth, and those paid $20.00 were not persuaded by the fact that the experimenter had told them that the task was enjoyable.

Although the interpretation of the results gets a little complicated, the explanation can be derived from the theory of cognitive dissonance. Basically, those students who were paid nothing and who had not been exposed to persuasion felt comfortable in telling the truth about how boring the experiment had been. Those who had been paid $1.00 felt uncomfortable about this, because they had acted in a way that was inconsistent with their real beliefs. They therefore changed their beliefs to be consonant with their actual behaviour. Those students who had been paid $20.00 were more prepared to see the payment as pressure and therefore that their behaviour was consistent with the amount of that pressure.

While there has been much argument about the best theoretical explanation for the results of this study, there is a substantial body of evidence to show that there is a reciprocal relationship between attitudes and behaviour (Eagly and Chaiken, 1993). This has important implications for attitude change. As was suggested earlier, most attempts to induce attitude change through the mass media are designed to effect a behavioural change in only a small percentage of the population. The idea is that a few per cent of people changing brand loyalty, for example, is sufficient to produce a significant profit for a cigarette company. A discussion of how to induce change in targeted individuals occurs in Section 7.3 of this chapter. Before proceeding to that discussion, it is useful to have some insight into why individuals resist attitude change.

7.2.3. Resistance to attitude change

If an attempt is to be made to change the behaviour of specific individuals, it is necessary to take into account explicitly those factors which may influence the change process. While there are many factors that may influence the change process, a broad understanding of the principles governing such factors can be derived from Festinger's theory of cognitive dissonance.

Festinger's (1957) theory of cognitive dissonance has been applied to the development of an explanation for people's resistance to attitude change. It is normal for us to be exposed to a variety of information, some of which is consonant with our existing beliefs and some of which is clearly dissonant. If we are to understand how attitudes are changed and what the sources of resistance to change are, it is necessary to explore the ways in which we manage, psychologically, discrepant or conflicting information.

Take, for example, the behaviour of a person who works in the pig industry. When entering the industry for the first time, the person seeks out all the information relevant to the job. This includes a variety of things, including techniques for feeding, assisting with matings, health management, handling techniques, and so on. This information is then combined on the basis of expectancy-value, that is, on the basis of the sum of the values of each attitude or belief relevant to a particular behaviour multiplied by how positively or negatively each behaviour is valued. The values of each behaviour are determined by feedback from workmates, behaviour of the animals, supervisors' comments and so on. Once a behavioural strategy is established, the person often engages in less rational behaviour to defend the action.

There are many specific strategies that an individual can use to resist attitude and behavioural change. While Festinger's (1957) theory offers a general insight into how these strategies work, it is useful to look at some of the specific ways in which individuals resist change. A knowledge of these specific strategies can help when designing a training procedure for experienced stockpersons, because such training does not simply involve the imparting of information, but involves the changing of established habitual attitudes and behaviours.

One method of defending an established behaviour is to attempt to reduce the dissonance that comes from new information. Using the example from the pig industry, if the stockperson is told that even moderate negative behaviours affect pig handling and production, the stockperson may say, 'Some pigs are difficult to handle and you need to use a moderate amount of force.' Here the person is reducing the importance of dissonant elements.

Another strategy to reduce dissonance is selective exposure. Selective exposure occurs when a person seeks out information that supports an earlier decision and avoids information that is dissonant. For example, a person may refuse to discuss the effects of handling on pig reproductive performance. In this way only new information supportive of the previous decision is processed.

Resistance to attitude change can also arise from what is termed 'inoculation' (McGuire, 1964). This refers to the process in which a person has been warned of the opposing view and therefore has a prepared defence against it. Thus a trainer may say to a stockperson, 'When you go back into the piggery, some of your workmates will ridicule the things you have learned. In fact, what you have learned are some important husbandry techniques and your workmates will just be talking out of ignorance.' In this way, the stockperson has become inoculated against information counter to his/her newly acquired beliefs and is therefore more resistant to change.

How strong the current attitude is will affect the amount of resistance to attitude change. If a person has a strong personal investment in a set of beliefs, these beliefs will be resistant to change. For example, a stockperson may strongly believe that it is essential that farm animals be taught who is boss because otherwise they will tend to dominate the stockperson and be difficult or even dangerous to handle. This person, therefore, may resist attempts to persuade him or her that fewer negative behaviours are necessary.

Some personality variables may contribute to resistance to attitude change. The authoritarian personality discussed earlier (Section 4.4) is an example of this. The work of Adorno *et al.* (1950) on the authoritarian personality showed that authoritarian people were reluctant to accept information from a person who was not seen to be an expert or to have a high degree of authority. This was later extended by Rokeach (1960) to include dogmatism to describe people who hold very strong opinions, both positive and negative, who have a strong trust in authority and who are intolerant of disagreement. Thus a stockperson high on dogmatism may resist attitude change because the trainer 'has never worked in a piggery' and is therefore not an expert.

Finally, some characteristics of the message and its source can affect resistance to attitude change. A major research programme conducted at Yale University in the 1950s and early 1960s attempted to identify these factors (Janis and Hovland, 1959). These factors are summarized in Fig. 7.1.

When attempting to change the attitudes and behaviour of individuals, it is necessary to take into account all of these sources of resistance to change. We shall discuss this in some detail in the next section.

7.3. CHANGING ATTITUDES AND BEHAVIOUR

As was discussed in Section 7.2.2, advertising campaigns through the mass media are designed to induce change in only a very small percentage of the population. Particular individuals are not the targets of such campaigns, because these campaigns work on the principle that a small percentage of individuals will be influenced by the campaign. When it comes to inducing behavioural change in individuals, where higher resources required need to

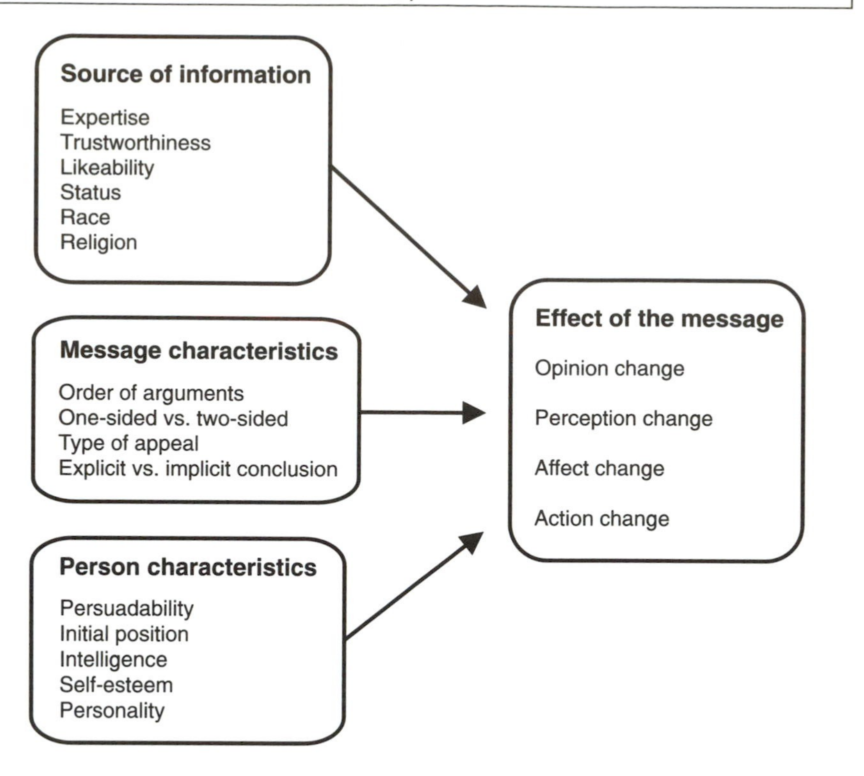

Fig. 7.1. Factors that affect attitude change.

be offset by a high success rate, techniques that merely deliver a message on a 'take-it-or-leave-it' basis are not appropriate.

7.3.1. Individual change

Inducing behavioural change involves processes somewhat different from those in the normal classroom situation. Typically, in a classroom situation, the objective is to impart knowledge. This is usually accomplished by providing information in a variety of interesting ways and by reinforcing the knowledge acquisition process, using various motivational, assessment and feedback techniques. In the classroom situation, no particular behavioural end-point is used; students are merely expected to develop their conceptual and analytic skills and knowledge.

Inducing behavioural change not only involves imparting knowledge and skills, but also involves changing established habits, altering well-established attitudes and beliefs and preparing the person to handle reactions from others towards the individual following change. In other words, the process of inducing behavioural change is really a comprehensive procedure, in which all of the personal and external factors that are relevant to the behavioural situation, which were outlined in the previous section, are explicitly targeted.

Of necessity, the costs involved in such a process are substantially higher per individual than those in classroom techniques, but the magnitude of change can be expected to be much larger than that induced by classroom teaching.

Cognitive–behavioural intervention techniques offer a powerful tool for changing attitudes and behaviour in individuals with a high success rate.

7.3.2. Cognitive–behavioural intervention

Cognitive–behavioural interventions are based on the idea that people have schemata about a particular set of objects. For example, a person working in the pig industry may have a particular set of beliefs about pigs, including 'they squeal, they have a distinctive smell, they are noisy'. These schemata also include affective components, that is, feelings about the objects that make up the schemata. Much of the cognitive–behavioural research has focused on treatment of people with inappropriate schemata. For example, people who inappropriately judge themselves as incompetent or inferior may not be able to cope with routine work or family problems, may underperform and may avoid activities for which they feel inadequate. Turk and Salovey (1985a, b) argue that schemata induce people to selectively attend to the world around them. This means that such individuals may behave inappropriately and that the schemata will be difficult to change.

However, attitude systems are examples of schemata, and normal attitude–behaviour relationships fit the cognitive–behavioural model. As will be discussed later, cognitive–behavioural techniques are very useful in changing attitudes and behaviour in agricultural industries.

Kendall and Hollon (1979) identified two basic principles underlying cognitive–behavioural techniques. The first principle was that cognitions are subject to the same laws of learning as are overt behaviours. The second principle is that attitudes, beliefs, expectancies, attributions and other cognitive activities are central to producing behaviour. These principles are quite consistent with all of our earlier discussions about the acquisition of attitudes and behaviour and about the complexity of factors that directly or indirectly influence the attitude–behaviour relationship.

Table 7.1. Cognitive–behavioural interventions in two situations.

Situation	Target	Approach
Behavioural pathology	Fear of snakes	Information provision Modify inappropriate cognitions Systematic desensitization
Stockperson behaviour	Inappropriate human–animal interactions	Provide factual information Modify inappropriate beliefs Animal-handling skills training

Although cognitive–behavioural techniques were developed to deal with individuals suffering from various behavioural pathologies – for example, extreme fear of spiders or snakes – the principles apply to non-pathological situations. As can be seen in Table 7.1, the use of cognitive–behavioural strategies appears to be entirely appropriate for normal situations where cognitive and behavioural change is required.

7.4. MODIFYING STOCKPERSON ATTITUDES AND BEHAVIOUR

Two studies have successfully applied cognitive–behavioural intervention techniques to improve stockperson attitudes and behaviour (Hemsworth *et al.*, 1994a; G.J. Coleman, P.H. Hemsworth and M. Hay, unpublished data) and these will be described in detail in this section to illustrate the techniques and their implication in agriculture.

7.4.1. Modifying the attitudes and behaviour of stockpeople at small independent farms

In the first study (Hemsworth *et al.*, 1994a), the cognitive–behavioural modification techniques involved retraining in behavioural areas as well as changing the attitudes and beliefs of people. Because of the reciprocal relationship between the attitudes and behaviour of the stockperson and the equally strong relationships between the stockperson's attitude and behaviour and the fear and reproductive performance of pigs (Hemsworth *et al.*, 1989), the cognitive–behavioural intervention procedure targeted both the attitudes and the behaviour of stockpeople. This was achieved by aiming firstly to improve the stockperson's beliefs about pigs, particularly beliefs about handling pigs, by providing stockpeople with key information on commercial pigs, such as the ease with which they can and should be handled, their sensitivity to the range of negative behaviours used by stockpeople (and their sensitivity to stressors in general) and the adverse effects of these negative behaviours on their fear of humans, which in turn can have negative consequences for their productivity and ease of handling. The indirect effects of negative handling on the stockperson, such as a deterioration in job satisfaction via difficulty in handling and closely inspecting fearful pigs, poor productivity and thus profitability and depressed welfare in fearful pigs, were also raised with the stockpeople. The training also provided stockpeople with information on the positive behaviours that can be used to reduce fear of humans.

Second, in order to address the behavioural aspects of the intervention, stockpeople were shown video footage of the behaviour of stockpeople in commercial units and emphasis was placed on those patterns, such as a high percentage of negative interactions, including moderate interactions, that have been shown to increase the pigs' fear of humans. Video footage of the behavioural responses of pigs to a range of stockperson behavioural patterns was also presented to assist stockpeople in recognizing and assessing fear responses in their animals.

Encouraging stockpersons to practise, immediately upon their return to the piggery, actual handling techniques in a variety of situations was an essential part of this process. Behavioural modification is necessary to ensure that there is a consonant change in beliefs and behaviour, so that the reciprocal relationship between these elements is maintained. In order to reinforce the information targeting improvements in both beliefs and behaviour, stockpeople were provided with written material, in the form of a booklet, posters and a regular newsletter. The desired outcome of the study was to reduce the percentage of negative interactions used by stockpeople on commercial pigs (i.e. reduce the degree of aversiveness of their behaviour towards pigs), and this cognitive–behavioural intervention procedure, which incorporated the above principles, was used.

Thirty-five commercial pig farms in Victoria and New South Wales, Australia, were selected. The farms varied in size from 75 to 300 breeding females and all breeding females were housed indoors on concrete, either in stalls or in groups during pregnancy and in farrowing crates during lactation. At each of these farms, one stockperson was predominantly responsible for conducting oestrus detection and supervising and assisting matings and this stockperson was the subject of study at these farms. In order to study the effects of the cognitive–behavioural modification treatment, there were premodification and postmodification periods at each farm, consisting of a minimum of 8 and 15 months, respectively. At the commencement of the premodification period and the termination of the postmodification period, observations and measurements were conducted on the attitude and behaviour of the stockperson at each farm responsible for mating activities, and observations were conducted at each farm on the approach behaviour of recently mated gilts and sows to an experimenter (to assess level of fear of humans) in the standard test discussed in Section 5.4. Reproductive records for each farm were collected monthly during the two periods. At the end of the premodification period, farms were randomly allocated, within level-of-fear strata, to two treatments, modification and control.

The modification treatment involved a 1-h individual session with each stockperson, in which the cognitive–behavioural intervention procedure was used. During this session, stockpeople were first shown the evidence for a relationship between stockperson attitude, stockperson behaviour, pig behaviour and pig productivity. This part of the procedure was designed to cover the cognitive aspects of the intervention. This evidence was derived from the on-farm studies and the handling experiments conducted by Hemsworth and colleagues (for example, Hemsworth *et al.*, 1981a, b, 1986b, 1989). Graphs of the results from these studies and the economic implications of the differences in productivity that could be attributed to stockperson behaviour were discussed. Furthermore, the importance of minimizing aversive behaviour to promote ease of movement in the pig and the influence of physical features in the environment on pig movement were outlined. Stockpeople were given an opportunity to question and discuss the evidence and, if doubts or uncertainty were expressed, the relevant parts of the evidence were reviewed.

Following this information session, stockpeople were shown video footage of positive and negative (aversive) behaviours displayed by stockpeople, and emphasis was placed on the effect that these behaviours would have on conditioning the pig's behaviour. This part of the procedure was designed to cover behavioural aspects of the intervention. It was emphasized that continual, even mildly, negative behaviour towards the pig would tend to condition the pig to be fearful of the stockperson. An 'inoculation' procedure (Section 7.2.3) was used, in which stockpeople were told that some negative behaviour might be necessary in some situations (e.g. a wary animal is reluctant to move out of a familiar area to an unfamiliar one) but that most of the negative behaviours in the industry were often avoidable and, in any event, should be greatly outnumbered by positive interactions with the pig.

In order to reinforce the information presented in this session, stockpeople were provided with posters to place in working areas of the piggeries. Also, regular newsletters were used to summarize the important points of the cognitive–behavioural intervention programme and to prompt an assessment by the stockperson as to whether or not changes in his or her behaviour and the behaviour of his or her pigs were being achieved.

The control treatment involved two observers, the same two who imposed the cognitive–behavioural intervention procedure, visiting the farm for a 1-h session, in which only general developments in the pig industry were discussed.

Stockperson attitudes and behaviour were assessed in the same way as in our earlier studies (see Chapters 4 and 5) and the behavioural response of pigs to humans was assessed in a standard test, in which the approach behaviour of the pig to a stationary experimenter in an arena was measured (see Chapter 5).

Relative to the control treatment, the modification treatment resulted in a greater increase in positive attitudes towards petting and talking to pigs (Table 7.2). Corresponding with this relative improvement in the attitude of stockpeople at the modification farms was a significant reduction in the percentage of physical interactions displayed by the stockperson that were negative in nature.

These relative improvements in the attitude and behaviour of stockpeople at the modification farms corresponded with a significant relative reduction in the level of fear of humans by pigs at these farms (Table 7.2). There were increases in the time that pigs spent within 0.5 m of the experimenter and in the number of interactions with the experimenter in the standard test for pigs at the modification farms relative to pigs at the control farms. Furthermore, there was a strong tendency for an increase in the

Table 7.2. Summary of the effects of a cognitive–behavioural intervention procedure, applied at modification farms, targeting the attitudes and behaviour of stockpeople in the pig industry (from Hemsworth *et al.*, 1994c).

Variables	Control farms	Modification Farms
Human attitudes*		
Beliefs about petting and talking	89.2	102.9
Beliefs about effort needed to handle	89.8	92.2
Human behaviour		
Negative behaviours (%)	55.8	38.6
Fear levels		
Time spent near experimenter (s)	15.6	21.9
Number of interactions with experimenter	1.3	2.0
Reproductive performance		
Piglets born per sow per year	22.2	23.8

* High score indicates positive beliefs.

number of pigs born per sow per year at the modification farms relative to the control farms (7% increase in reproductive performance at the former group of farms; Table 7.2).

The intervention procedure used in this study was less intensive than that normally used in therapeutic environments, where a number of sessions over an extended period of time are used and where the behavioural component would normally include actual behaviour sessions in which the person used role-playing or even actual situations to learn the new behaviour patterns. Ideally, interventions in this area should be conducted in a place where opportunities for behavioural practice are available and follow-up sessions should be used. The resources available for the research presented here did not permit such a comprehensive programme. Nevertheless, despite these shortcomings, behavioural and attitudinal improvements were observed following intervention and the effects of these were reflected in changes in pig behaviour. This suggests that interventions of this kind are a powerful tool for improving the productivity and welfare of commercial pigs.

The effects of the intervention on the reproductive performance of pigs, while in the expected direction, were not substantial in magnitude. This raises the issue of whether the results from this study provide support for the sequential model, described in Chapter 6, in which stockperson attitudes affect stockperson behaviour, which, in turn, affects pig behaviour and performance. Had the results only shown an effect on stockperson attitudes and behaviour, no firm conclusions could be drawn, because such results could have been attributed to stockpeople merely conforming to the expectations of the experimenters. However, the observed reductions in the level of fear of humans by the pigs at the modification farms cannot be readily explained other than by the effects of changed stockperson behaviour. This, in conjunction with the small but marked improvement in reproductive performance of pigs at the modification farms (7% relative to the control farms), strongly suggests a causal link between stockperson attitudes and behaviour, on the one hand, and pig behaviour and performance, on the other. Fear–reproduction relationships observed in the industry and the effects of handling studies on the growth and reproductive performance of pigs further support this interpretation.

7.4.2. Modifying the attitudes and behaviour of stockpeople at a large integrated farm

A second study (G.J. Coleman, P.H. Hemsworth and M. Hay, unpublished data) using cognitive–behaviour modification techniques was conducted in a commercial piggery where large numbers of stockpeople worked together rather than on individual farms. The procedure used in this study was similar to that used in the previous study. However, a number of refinements were made to improve both the efficacy of the procedure and the practicability of the procedure for training stockpeople in a formal group setting. Again, this procedure was designed, firstly, to modify the beliefs of stockpeople about the sensitivity of pigs, handling pigs and the consequences of aversive handling on the ease of handling and productivity of pigs and, secondly, to educate stockpeople to properly observe and handle pigs to avoid these adverse handling effects. An important component of this training procedure was a detailed series of reviews and discussions of the subject, utilizing data and observations from our earlier studies, emphasizing the important relationships between stockperson attitude, stockperson and pig behaviour and pig stress and productivity. In addition, a booklet summarizing the procedure and specialized video footage, showing examples of appropriate and inappropriate behaviours by stockpeople and the accompanying behavioural responses by pigs, were presented. This video and written material was also used for revision by the stockpeople.

This intervention procedure was conducted on half of the stockpeople 1 month after the study commenced and the remaining stockpeople were subjected to a control procedure. At the commencement of the study (i.e. premodification for all stockpeople) and then a month later, the attitude and behaviour of all stockpeople and the behaviour of pigs assisted to mate by these stockpeople were assessed. Stockperson attitudes and behaviour were assessed in the same way as had been done in all previous studies. In this study, however, an attempt was made to assess the effects of the cognitive–behavioural interventions on other job-related variables as well. The questionnaire developed by Coleman *et al.* (1998) and described in the previous chapter (Section 6.3) was used to measure a number of job-related measures, such as job satisfaction and job knowledge.

In most of our previous research, the behavioural response of pigs to humans was assessed in the standard test described in Chapter 5. It was not possible to conduct this test, due to the amount of time needed and the costs of setting up specific arenas at various locations. A simpler test, which has been shown to be predictive of the above test in a commercial setting (Hemsworth *et al.*, 1981b), was used in this study and basically involved measuring the withdrawal response of feeding gilts and sows to an experimenter approaching in a standard manner. These animals were housed in stalls and 30 pigs supervised and assisted to mate by each stockperson were tested 3 weeks postmating. Because, early in the study, stockpeople had difficulties recording the identities of pigs which they had assisted to mate, insufficient data were collected in the premodification period to obtain reliable data from each stockperson. Nevertheless, the data were collated and analysed, and are presented with the qualification that the premodification data are unreliable because of the small numbers of animals mated by each stockperson.

Consistent with results from the previous study (Hemsworth *et al.*, 1994a), stockperson attitudes improved following the training procedure (Table 7.3). In the modification group, mean responses on attitude towards working with pigs increased, indicating that attitudes improved, while those for the control group actually decreased. Similarly, mean scores on beliefs about petting and talking to pigs increased for the trained stockpeople, indicating an improved attitude, while those for the control group decreased. Scores for beliefs about handling non-oestrous pigs increased in the modification group compared with the control group, also indicating an improvement in attitudes in the trained stockpeople. No significant changes were observed in job-related measures, although there was a tendency for stockpeople in the modification group to report increased knowledge compared with those in the control group. It is interesting to note that all the attitudes showing improvement were in the area of behaviour. Given that the

Table 7.3. Summary of the effects of a cognitive-behavioural intervention procedure, applied at a large commercial farm, targeting the attitudes and behaviour of stockpeople (from G.J. Coleman, P.H. Hemsworth and M. Hay, unpublished data).

	Percentage change from the pre- to postmodification period	
Variables	Control group	Modification group
Human attitudes*		
Beliefs about petting and talking	–6%	+6%
Beliefs about effort needed to handle	–14%	+8%
Human behaviour		
Negative behaviours (%)	–17%	–22%

* High score indicates positive beliefs.

aim of the training programme was to change stockperson behavioural beliefs, this was a desirable outcome. As in the previous study, it is possible that changes in attitude score could reflect the fact that stockpeople in the modification group were aware of what was required and simply answered in a way that was consistent with the experimenters' expectations.

However, stockperson behaviour in the modification group, which was assessed by different experimenters from those who supervised the attitude assessment, also showed a decrease following the training programme in both number and percentage of negative behaviours towards pigs, relative to the control group. Given that stockpeople were not aware of the way in which their behaviour was being assessed, these changes in observed behaviour suggest that a genuine change in stockperson attitudes and behaviour had occurred. Furthermore, there was a strong trend, consistent with this improvement in stockperson behaviour, for pigs supervised and assisted to mate by stockpeople in the modification group to show less withdrawal to an approaching experimenter. No such effect was evident in the control group. This indicates that, despite the fact that the gilts and sows in this large commercial farm had been exposed to many stockpeople, there was some influence on their behaviour exerted by the stockperson who was involved in mating the sows. In the earlier study (Hemsworth *et al.*, 1994a), one stockperson was predominantly responsible for supervising and assisting mating activities at each farm. Therefore, consistent with results from the previous research, stockperson attitudes, stockperson behaviour and, to a lesser extent, pig behaviour improved following the training procedure.

It is important to appreciate that regular handling of a predominantly positive nature is required to reduce fear responses in pigs. Furthermore, the period required to reduce fear levels in those animals experiencing high levels of fear will be considerable, perhaps months (Hemsworth *et al.*, 1981a). While changes in fear levels may be observed in the short term, a greater period of time may be required before stress responses, either acute or chronic, in those highly fearful animals are reduced to the extent where reproductive performance is not limited. It was not possible to monitor changes in the long term in this study, as in the earlier study (Hemsworth *et al.*, 1994a), because stockpeople were regularly moved to different areas of production at the commercial farm.

The results of this study confirm that stockperson attitudes and behaviour can be improved in a large commercial farm and that short-term effects on pig behaviour can be observed. Given the peer pressure operating on individual stockpeople by workmates in the same unit and the pressures that the unit manager would exert to produce conforming behaviour by stockpeople, this is a very encouraging result.

An interesting result which emerged from this study and which was not anticipated when the study was commenced related to employee turnover. In the 12 months following the study, the group that had received training showed a 52% higher retention rate than did the stockpeople who received no intervention. This suggests that one effect of the modification procedure was to improve long-term job satisfaction, even though no clear changes in job satisfaction were observed just 1 month after the modification procedure.

7.5. REVIEW – MODIFYING STOCKPERSON BEHAVIOUR

The results from the two intervention studies reported in this chapter, taken in conjunction with previous research on the sequential relationship between stockperson attitudes, stockperson behaviour, pig behaviour and pig productivity and research on handling pigs, indicate an important role for training stockpeople. Indeed, taken in conjunction with this earlier research, there is a strong case for introducing stockperson training courses in the pig industry that target the attitudes and behaviour of the stockperson. These two intervention studies also demonstrate that this training programme is both practical and effective with a wide range of pig stockpeople working in a variety of situations. Although there is substantially less information available about other intensive-farming industries, it is likely that these other industries would also benefit from such training programmes.

In Chapter 2, we discussed at some length the need for stockpeople to be treated as professionals and to receive due recognition for the central role that they play in animal productivity and welfare in situations where animals are intensively farmed. Training stockpeople to have a sound factual knowledge base about human and animal behaviour and to clearly understand and be able to apply handling techniques offers an important step towards meeting this need.

In the next chapter, we shall review the material covered in this book and look towards the future for people working in the intensive-farming industries. In particular, we shall propose ways in which the research we have reviewed can be applied to industry for both selection and training of stockpeople.

Chapter 8

Conclusion: Current and Future Opportunities to Improve Human–Animal Interactions in Livestock Production

8.1. INTRODUCTION

We began this book by looking at the relationship between humans and domestic animals. In doing so, we looked at the welfare issues associated with our use of animals in agriculture. This was a key place to begin the book. Any discussion of the role of domestic animals in our society needs to take into account the welfare of animals and our obligation towards them with respect to welfare. We then turned our attention to the need to optimize animal productivity and welfare and the place of the professional stockperson in the livestock industries. This then led to an account of human–animal interactions. This first section of the book was designed to set the stage for a more detailed account of our empirical knowledge of human–animal interactions and their consequences for agricultural industries and the theory underlying this empirical research.

Human–animal interactions have been the topic of this book because there is an increasing body of evidence, currently not widely recognized in agriculture, that suggests that these interactions may result in profound behavioural and physiological changes in the animal, with consequences for the animal's performance and welfare. Furthermore, these interactions may also influence the stockperson to the extent that job-related characteristics, such as job satisfaction, motivation and commitment, may be affected, with implications for the job performance and career prospects of the stockperson.

Understanding stockperson behaviour appears to be the key to manipulating these human–animal interactions to improve animal performance and welfare, as well as the stockperson's attitude and motivation regarding the job. An understanding of stockperson behaviour is necessary if we wish to influence or change stockperson behaviour. As discussed in Chapter 4, the theory of reasoned action as proposed by Fishbein and Azjen (1974), implies that behavioural change is ultimately the result of changes in beliefs. This, in turn, implies that in order to influence behaviour, we have to expose stockpeople to information that will produce changes in their beliefs. Alternatively, we may wish to select stockpeople to work in the animal industries, at least in part, on the basis of their beliefs and thus behaviour towards farm animals. The ability to manipulate these human–animal interactions, via training stockpeople in terms of their attitudes and behaviour towards their animals and/or selecting stockpeople on the basis of these characteristics, has the potential to markedly and quickly influence animal welfare and performance. Another option is to completely automate the stockperson's functions and thus avoid the deleterious effects of adverse human–animal interactions. With the high cost of labour in many Western countries, this option may appear attractive to the livestock industries. However, the trade-offs in terms of not only costs but also effects on rural society would need to be carefully considered. Also, automated systems for all aspects of animal husbandry may not be practicable. For example, routine checking of health and welfare, moving animals from one place to another as part of normal farm practice and introducing animals or removing animals from the farm may require some direct human intervention. Once this happens, there is an opportunity for the stockperson to influence animal productivity and welfare. This chapter will examine the opportunities to improve not only the stockperson's attitude and behaviour towards his or her animals, but also to improve the stockperson's attitude and motivation to the job in order to improve human–animal interactions and animal performance and welfare.

8.2. THE RELATIVE IMPORTANCE OF THE STOCKPERSON IN FARM PRODUCTION

The focus of production research and, recently, welfare research has been on factors such as the housing, nutrition, breeding, health and climate. The clear justification for this research has been that paying attention to these factors will substantially improve the economics of farm production and the quality of the farm product. While these factors clearly affect farm production and, in fact, animal welfare, this focus has often been at the expense of recognizing and investigating the role of the stockperson. Research examining fear–productivity relationships in the dairy, pig and poultry industries has consistently shown that fear of humans is one of the single most important factors associated with variation between commercial farms in animal productivity (Chapter 6, Section 6.2.3).

In Chapters 1 and 7, we discussed in detail the need for stockpeople to be treated as professionals and to receive due recognition for the central role that they play in animal productivity and welfare. It was emphasized that training stockpeople to have a sound factual knowledge base about human and animal behaviour and to clearly understand and be able to apply handling techniques offers an important step towards meeting this need. Furthermore, the opportunity to undergo training is likely not only to influence the skill and knowledge base of the stockperson, but also to improve the self-esteem, job satisfaction and work motivation of the stockperson, with possible advantages to work performance and retention rates. It is not difficult to appreciate that, in addition to direct effects on fear and, therefore, effects on the productivity and welfare of the animals, improvements in the stockperson's attitude and behaviour towards the animals may affect these other job-related characteristics through improvements in ease of handling, ability to closely supervise and assist or address issues when necessary and gains in animal performance and welfare.

Limited research has provided some evidence for the association of personality traits of the stockperson with animal productivity. For example, there is evidence in both the dairy and pig industries that certain personality traits may be associated with animal productivity. However, the causal basis of these relationships and the opportunities to manipulate any causal relationships need to be examined. Whether personality factors operate indirectly by influencing the formation and maintenance of stockperson attitudes or whether these factors have a direct effect on other aspects of the stockperson's job performance is yet to be determined. Nevertheless, selection procedures may offer opportunities to improve farm productivity if a causal relationship between stockperson personality and farm productivity can be identified; an understanding of personality factors may be useful in matching people to some types of jobs in agriculture. Since work motivation and job commitment are influential in affecting job performance, there may also be opportunities to select stockpeople on the basis of their predicted work motivation and job commitment.

However, before exploring these aspects in too much detail, it is relevant to consider, as an alternative, the possibility of reducing or eliminating the role of the stockperson in animal production entirely. This is the substance of the next section.

8.3. THE NECESSITY OF HUMAN–ANIMAL INTERACTIONS

The necessity of human contact in animal-production systems has been reviewed by Hemsworth and Gonyou (1997) and it is useful to use this previous discussion as a basis for a discussion on the possibility of eliminating all human–animal interactions in animal-production systems in order to eliminate the chances that these interactions may limit animal performance and welfare.

Modern animal production involves several levels of interaction between stockpeople and their animals. Many interactions are associated with regular observation of the animals and their conditions and thus this type of interaction often involves only visual contact between the stockperson and the animals, perhaps without the stockperson entering the animals' pen. Visual and auditory interactions may also be used to move animals and, in some industries, tactile interactions may be used. All interactions contribute to the overall relationship that animals have with humans and determine if that relationship is positive, neutral or negative. Because of the potential for negative interactions by the stockperson, Hemsworth and Gonyou (1997) have questioned the necessity of interactions between humans and animals. To examine this question, it is necessary to consider the human–animal interactions in agriculture in more detail.

An essential role of stockpeople in achieving high animal performance and welfare is the careful observation of animals under their care. Although animal conditions such as ambient temperature, noxious gas levels, presence of feed and water can be remotely monitored, direct observation of animals often provides the first evidence of departure from normality in animals. In particular, behavioural change can be utilized by stockpeople to identify abnormality, such as illness or stress. For example, an animal not feeding, a social animal voluntarily separated from the group or an animal unresponsive to environmental change, such as the approach of the stockperson, can be used to identify the early stages of a problem and enable a prompt response to the problem. Often the diagnosis of the problem by veterinarians relies on reports from the stockperson on the behaviour of the animal. For example, signs of pacing and kicking at its belly, are indicative of colic in horses. One could argue that the prompt identification of an impending problem for the animal relies heavily on the observation of the stockperson on animal behaviour. Use of video cameras is in general less effective than direct observation. For example, video cameras are limited by their inability to use localized non-visual cues, such as auditory ones (vocalizations), and to observe the fine detail that is necessary to discriminate, for example, between a lesion and a smear of dirt on the skin. While animal movement can be quantified with sophisticated tracking systems, a human observer is still necessary in many cases to make a decision on whether or not a video of an unusual behavioural pattern in an animal is indicative of a problem. Clearly, automation should be utilized to assist the stockperson in monitoring animals and their conditions and, in fact, automation of the tedious, laborious and objectionable tasks, such as cleaning, provision of feed and water, etc., which may decrease job satisfaction, should be encouraged. However, automation is unlikely, in the foreseeable future, to completely replace the stockperson. In the interests of animal welfare alone, the general public probably considers careful observation of animals under the stockperson's care as an essential part of good stockmanship. While not necessarily a legal requirement, codes of practice for farm animals in many countries often state that daily observation of animals in confined conditions is essential.

Animals in most production systems have to be moved. Extensively grazed animals, such as beef cattle and sheep, are moved between pastures as part of optimal grass management and extensively grazed dairy cows are moved several times a day during lactation to be milked. Growing pigs are generally moved from pen to pen, in order to provide accommodation suitable to their stage of growth, and breeding pigs may be regularly moved according to their stage of the breeding cycle. It is during these situations that human–animal interactions have considerable potential to influence animal performance and welfare. As considered earlier in the book (Chapters 5 and 6), the risk of eliciting high fear of humans can be reduced by stockpeople minimizing their negative interactions while maximizing their positive ones. This can be achieved by using negative interactions, such as hits and slaps on dairy cattle and pigs, only when necessary and seeking opportunities to use positive interactions, such as pats, strokes and the hand of the stockperson resting on the animal when animals are moving, have arrived at the destination or are feeding (Fig. 8.1). However, humans inadvertently interact with animals when they inspect animals and equipment, such as feeders and waterers in their pens. Although these observations do not necessarily involve contact with the animals, these visual interactions can also be potentially fear-provoking. The avoidance of fast speed of movement and unexpected movement or appearance by the stockperson will assist in reducing fear of humans by poultry. Recent unpublished observations by the authors on commercial dairy cows also indicate that visual interactions, such as waving, and auditory contact, such as shouting, may also be fear-provoking.

Human–animal interactions also occur in situations in which animals must be restrained and subjected to management or health procedures. Some animals may never be restrained during their lives, while others are restrained on a regular basis. Some sort of restraint is used for weighing, milking, vaccinating and blood sampling, and animals are restrained for procedures that are probably painful, such as castration, branding, ear-tagging and dehorning. It may be possible to reduce or eliminate some of these procedures. Procedures such as vaccinations and blood sampling for diagnosis are necessary to improve the health and thus welfare of the animals and some degree of discomfort or pain is justified in the welfare interests of the animal. Procedures such as milking and shearing are directly related to the reason the animals are kept and could only be eliminated if the industry is banned. Weighing, ear-tagging, castration and dehorning are justified by facilitating management, improving product quality or reducing the possibility of injury to animals or humans. Although some reduction in these procedures may be possible, it is likely that they will remain part of animal care and production for some time.

The association of fear and pain from these husbandry procedures with the humans performing them will increase the fear of humans that animals exhibit in other situations, such as during routine inspections. The effect these procedures have on the human–animal relationship relates both to the aversiveness of the procedure and the association of people with that aversion. Rewarding experiences, such as provision of a preferred feed or even positive handling, around the time of the procedure, may ameliorate the aversiveness of the procedure and reduce the chances that animals associate the punishment of the procedure with humans. For example, studies with pigs have shown that pigs will associate the rewarding elements of feeding with humans if handlers are present at feeding (Hemsworth *et al.*, 1996c). Hutson (1985) found that, although the effectiveness of food rewards diminished as the severity of the handling treatment increased, rewarding sheep

Fig. 8.1. The frequent use of positive interactions by stockpeople, such as pats and the hand resting on the animal's back, will decrease the pig's fear of humans.

with barley food rewards improved subsequent ease of handling in the location in which the aversive treatment was previously imposed. Surprisingly, daily injections were not highly aversive to pigs (Hemsworth *et al*., 1996b), and the authors suggested that there may have been some rewarding elements for the pigs in these handling bouts, such as the presence of and opportunity to closely approach and interact with the handler before and after injection. Rushen and colleagues (Munksgaard *et al*., 1995: Rushen *et al*., 1995; de Passille *et al*., 1996) have shown that performing an aversive treatment at a specific location or by either an unfamiliar or familiar operator wearing different distinctive clothing may avoid dairy cows associating the procedure with the regular stockperson.

As Hemsworth and Gonyou (1997) have suggested, there are opportunities to reduce or even eliminate human involvement in some animal-management procedures that are aversive to the animal. Examples include robotic shearing of sheep, robotic milking of cows and automated handling facilities for sheep (Syme *et al*., 1981) and pigs (Barton Gade *et al*., 1992). The effect of eliminating humans from such handling procedures on animal responses is well illustrated by research on mechanical and manual harvesting of broiler chickens (Duncan *et al*., 1986). While maximum heart rates of birds caught by either method were similar, the rates remained high for longer in manually caught birds than in birds caught by a specially designed machine. In tonic immobility tests, manually caught birds showed a longer response, possibly indicative of greater fear. These results indicate that the stressfulness of some procedures may be reduced by eliminating humans from the procedure. Similar opportunities may exist with other management and health procedures, and clearly research on these procedures should be conducted to determine the effects of the component of the procedure involving human contact on the animals' responses. In situations where the human contact component is highly aversive or even injurious to the animal, procedures that eliminate human involvement or changes in the behaviour of the human should be sought. For instance, since the method of catching laying hens in cages affects the incidence of bird injuries (Gregory *et al*., 1993), catching techniques that minimize injury should be identified and adopted.

Another potential problem for animals that are deprived of human contact is the fact that, if human contact is required, perhaps in an emergency situation, this interaction will be highly fear-provoking and aversive. One could also question whether or not humans are viewed as social partners in the environment and whether or not they provide environmental variation, something that may be minimal in a totally automated system if not specifically addressed.

On balance, there appears to be a strong case for stockpeople to continue to interact with livestock. Even if an opposing case could be mounted, no cost–benefit analysis of eliminating the stockperson has been done and, given the capital cost of many alternatives, it is unlikely that such elimination would be practical. Therefore in order to maximize opportunities to benefit from the stockperson's input into farm production, the next two sections deal with managing this important resource in agriculture: training and selecting stockpeople to improve human–animal interactions.

8.4. TRAINING STOCKPEOPLE TO IMPROVE THEIR INTERACTIONS WITH THEIR ANIMALS

In the past, there have been limited opportunities for stockpersons to receive formal training on any aspect of their work. Typically, the new stockperson is given a brief orientation and is then placed in the work environment where he or she is expected to learn 'on the job'. On smaller independent farms, the owner/operator will usually perform this function. On larger farms, it is usually the immediate supervisor who will oversee this process, while coworkers provide day-to-day feedback. We would suggest that this practice reflects the belief that the stockperson has a set of mechanical functions to perform and, so long as these are done well, the farm will produce well. In general, where systematic training is employed, it usually targets management rather than stockpeople.

In Australia, for example, there have been many attempts to develop technical training courses suitable for stockpeople. However, in a recent attempt to set a standard for such training (Rural Training Council of Australia, 1990), a programme was developed for pig stockpeople that required not only the acquisition of a broad range of practical skills, but also included some fairly detailed technical and mathematical competencies and personal-effectiveness skills for stockpeople. One of the limiting factors for training stockpeople is that they usually have a restricted educational background, may not handle formal training particularly well and, in some cases, may have literacy problems. The fact that stockpeople may often come from such disadvantaged groups means that training needs to be carefully targeted and also needs to be delivered in a way that is accessible to such a group. Appropriate targeting means that the key skills that a stockperson requires must be identified and the training programmes that are developed must specifically relate to those skills. Accessible delivery means that the training programme should not be too formally presented, should not be too theoretical in content and should provide content to which stockpeople can easily relate.

An example of such a training programme is the one developed by the authors and described in Chapter 7. This programme, you will recall, was specifically designed to target those attitudes and behaviours of the stockperson that had a direct effect on pig behaviour and productivity. Training stockpeople by targeting for improvement both their key beliefs and behaviours that are influential in regulating human–animal interactions in a commercial setting offers the animal industries opportunities to improve animal performance and welfare. In the pig industry, these key human characteristics have been identified and an intervention procedure based on cognitive– behavioural principles has been shown to be effective in improving human–animal interactions in the pig industry.

Similar opportunities to improve human–animal interactions and thus animal performance and welfare may soon be available in other livestock industries.

8.5. SELECTING STOCKPEOPLE TO IMPROVE THEIR INTERACTIONS WITH THEIR ANIMALS

The possible benefits of selecting stockpeople appropriate to work with animals was discussed in Chapter 1. There is a wide variety of tests which focus on personality, vocational preference, work motivation, etc. The specific needs for selecting stockpeople, however, mean that many of these tests are not appropriate for the subject population involved. Further, there are no well-validated tests for empathy towards animals and no published tests at all for stockperson attitudes and beliefs. In choosing tests for use in selection of stockpeople, issues that need to be considered include the following:

1. Is the test designed as a pre-employment tool?
Tests that specifically refer to a current job or require specific knowledge of a particular agricultural industry are clearly inappropriate for selecting stockpeople to work in an agricultural industry for the first time.
2. Is the language appropriate for stockpeople?
Tests that use local vernacular or sophisticated terms may not be appropriate for a target population from which stockpeople are likely to come.
3. What is the length of the test?
If applicants are to fill out a number of tests, no one test should be unduly long. A total test battery that takes more than an hour or so to fill out is probably inappropriate for selecting stockpeople.
4. What does the test measure?
Care should be taken to ensure that the test measures variables that are appropriate for selection purposes. There are many published tests that may be useful for providing employees with advice, for identifying areas in which training might be appropriate or for characterizing the range of qualities that people in particular employment sectors might have. Such tests may not be appropriate tools for selection. Tests used for selection purposes must be able to discriminate with acceptable accuracy those people who will perform well in the job for which they are selected from those who will not perform well.
5. Is the answer format simple?
Tests chosen for use in selecting stockpeople may not give reliable results if some people are likely to have difficulty in responding to questions.

6. How is the test presented?
Given that there may be some literacy problems with the target population, the use of a multimedia format to present the tests may be appropriate.

In summary, procedures to select people to work as stockpeople need to take into account the particular characteristics of the target population as well as those characteristics that make a person well suited to work with animals in the livestock industries.

The authors' research has shown that there are aspects of stockperson attitudes and behaviour that affect productivity in intensive-farming industries. A test battery that targets those specific attitudes and behaviours as well as those generic characteristics that are predictive of work performance may well be appropriate. At present, there have been no research studies that have investigated the entire range of possible factors which may be relevant to stockperson performance. Where people have had previous experience working in a particular industry, our research has shown that their attitude towards working with animals in that industry is a good predictor of their behaviour and, ultimately, farm production. While other studies have suggested that personality may be relevant and that empathy may be important, these variables need to be studied further. As yet, no research has focused on other variables, such as attitudes to work, work motivation and work preference.

Because there is an increasing recognition of the need to employ people who will be adaptable and conscientious and who will treat animals well, a demand will develop for a selection procedure that can be widely used. No such procedures exist in agricultural industries at present.

While large organizations would be able to use selection procedures described above by utilizing their existing human-resources staff, smaller operators would need to have access to appropriate people to conduct selection on their behalf. A similar situation exists for training stockpeople on site. There appears to be a clear opportunity for training and selection to be provided for livestock industries by consultants.

8.6. THE BENEFITS FOR STOCKPEOPLE OF WORKING WITH ANIMALS

Working in farming industries offers an opportunity, from a dwindling range of opportunities, for people to obtain employment in rural areas. This facilitates people remaining with friends and relatives, living in a community with which they identify and having the financial base to live and raise a family in the district where they choose to live. While the better-educated youth from rural areas probably enter rural businesses or migrate to the city, those from rural areas with limited educational opportunities or achievements have less choice. Thus, there may be quite a strong motivation to succeed in jobs that require limited formal training.

In addition, while stockpeople may be dissatisfied with their job (English *et al.*, 1992), one can question whether most stockpeople actually dislike working with animals. Keeping pets is common in most families and, while the potential benefits are well recognized for children, such as promoting the development of social competency and responsibility (Edney, 1992), it is highly likely that adults gain considerable satisfaction and enjoyment from keeping pets. Surveys have shown that most people own pets for largely emotional reasons, which include companionship and to provide love and affection (Leslie *et al.*, 1994). It is also considered that pet ownership provides a range of other rewards for owners, such as support and a form of interest outside themselves (Edney, 1992).

Fig. 8.2. In addition to the responsibility and satisfaction derived from successfully caring for farm animals, stockpeople may also enjoy interacting with their livestock.

Interviews of several hundred stockpeople in the pig and dairy industries in Australia by the authors surprisingly indicate that, while many expressed a dislike for various aspects of the job, a clear majority of stockpeople (86% and 76% of pig and dairy stockpeople, respectively) enjoyed working with their animals (Fig. 8.2). Therefore, it is not unreasonable to suggest that working with animals provides stockpeople with a number of benefits, such as companionship and a commitment and interest that offers both responsibility and a sense of satisfaction for the health and welfare of lives other than those of themselves or their families. Indeed, successfully caring for farm animals may provide greater rewards for people than those gained by successfully working on a production line producing inanimate objects.

8.7. CONCLUSION

Human–animal interactions can have profound effects on farm animals. Regular negative interactions by stockpeople can result in the animals developing stimulus-specific fear responses to humans. As a result of stress responses, high levels of fear of humans can depress both the welfare and the performance of farm animals. Our knowledge of human–animal interactions is still limited, but several avenues for improving human–animal interactions to improve animal performance and welfare exist.

Interactions with animals may be either positive or negative in nature, and stockpeople should be aware that it is important not only to make all interactions as positive as possible, but also to ensure that the percentage of negative interactions is kept low. If some procedures involve negative interactions, it may be possible to eliminate the procedure altogether, accomplish the procedure mechanically and thus remove the human association, or to compensate for the negative interactions by additional positive interactions. Rewarding experiences at the time may also alleviate the aversiveness of the situation. Evidence in a number of livestock industries has shown that the stockperson's behaviour is ultimately a consequence of his or her beliefs about handling and interacting with farm animals. Therefore, in order to influence stockperson behaviour, stockpeople have to be exposed to information that will produce changes in their beliefs about handling and interacting with animals. Alternatively, we may wish to select stockpeople to work in the livestock industries on the basis of their beliefs and thus their behaviour towards farm animals. Research has shown that targeting these key attitudes and behaviour can indeed improve animal productivity, via reductions in fear and stress.

The human–animal relationship also has immediate and long-term implications for the stockperson. The human–animal interactions may affect the stockperson to the extent that job-related characteristics, such as job satisfaction, motivation and commitment, may be affected, with implications for the job performance and career prospects of the stockperson. This, in turn, is also likely to affect the performance and welfare of the livestock.

Therefore, selection and training procedures for stockpeople that target the attitude and behaviour of the stockperson offer a considerable opportunity to improve animal welfare and performance. This is the new direction for industries in which stockpeople regularly interact with livestock. Much has been done to improve genetics, nutrition, reproduction, health and housing, but efforts to target the stockperson, who performs such a key function, have just begun. We should not underestimate the role and impact of the stockperson with regard to animal performance and welfare. To do so will seriously risk the performance and welfare of our livestock. It is likely that both the livestock industries and the general community will place an increasing emphasis on ensuring the competency of stockpeople to manage our livestock; the livestock industries' interests are likely to be motivated by both animal productivity and welfare and the general community's interest in animal welfare.

References

Adams, B.M. (1968) Effect of cortisol on growth and uric acid excretion in the chick. *Journal of Endocrinology* 40, 145–151.

Adorno, T., Frenkel-Brunswik, E., Levinson, D. and Sanford, N. (1950) *The Authoritarian Personality*. Norton, New York.

Ajzen, I. (1988) *Attitudes, Personality, and Behaviour*. Open University Press, Milton Keynes.

Ajzen, I. and Fishbein, M. (1977) Attitude–behaviour relations: a theoretical analysis and review of empirical research. *Psychological Bulletin* 84(5), 888–918.

Ajzen, I. and Fishbein, M. (1980) *Understanding Attitudes and Predicting Social Behaviour.* Prentice-Hall, Englewood Cliffs, New Jersey.

Allport, G.W. (1935) Attitudes. In: Murchison, C. (ed.) *Handbook of Social Psychology*. Clark University Press, Worchester, Massachusetts, pp. 798–844.

Arave, C.W., Mickelsen, C.H. and Walters, J.L. (1985) Effect of early rearing experience on subsequent behavior and production of Holstein heifers. *Journal of Dairy Science* 68, 923–929.

Arave, C.W., Albright, J.L., Armstrong, D.V., Foster, W.W. and Larson, L.L. (1992) Effects of isolation of calves on growth, behavior and first lactation milk yield of production of Holstein heifers. *Journal of Dairy Science* 75, 3408–3415.

Aronson, E. (1969) The theory of cognitive dissonance: a current perspective. In: Berkowitz, L. (ed.) *Advances in Experimental Social Psychology,* Vol. 4. Academic Press, New York, pp. 1–34.

Asch, S.E. (1956) Studies of independence and conformity: I. A minority of one against a unanimous majority. *Psychological Monographs: General and Applied* 70, 547–549.

Bagozzi, R.P. (1980) *Causal Models in Marketing.* John Wiley & Sons, New York.

Barker, I.K., Beveridge, I., Bradley, A.J. and Lee, A.K. (1978) Observations on spontaneous stress-related mortality among males of the dasyurid marsupial *Antechinus stuartii* Macleay. *Australian Journal of Zoology* 26, 435–437.

Barnett, J.L. (1973) A stress response in *Antechinus stuartii* (Macleay). *Australian Journal of Zoology* 21, 501–513.

Barnett, J.L. and Hemsworth, P.H. (1990) The validity of physiological and behavioural measures of animal welfare. *Applied Animal Behaviour Science* 25, 177–178.

Barnett, J.L. and Hemsworth, P.H. (1991) The effects of individual and group housing on sexual behaviour and pregnancy in pigs. *Animal Reproduction Science* 25, 265–273.

Barnett, J.L. and Hutson, G.D. (1987) Objective assessment of welfare in the pig: contributions from physiology and behaviour. In: APSA Committee (ed.) *Manipulating Pig Production.* Australasian Pig Science Association, Werribee, Victoria, Australia, pp. 1–22.

Barnett, J.L., Hemsworth, P.H. and Hand, A.M. (1983) The effect of chronic stress on some blood parameters in the pig. *Applied Animal Ethology* 9, 273–277.

Barnett, J.L., Winfield, C.G., Cronin, G.M., Hemsworth, P.H. and Dewar, A.M. (1985) The effect of individual and group housing on behavioural and physiological responses related to the welfare of pregnant pigs. *Applied Animal Behaviour Science* 14, 149–161.

Barnett, J.L., Hemsworth, P.H. and Winfield, C.G. (1987) The effects of design of individual stalls on the social behaviour and physiological responses related to the welfare of pregnant pigs. *Applied Animal Behaviour Science* 18, 133–142.

Barnett, J.L., Hemsworth, P.H., Cronin, G.M., Winfield, C.G., McCallum, T.H. and Newman, E.H. (1988) The effects of genotype on physiological and behavioural responses related to the welfare of pregnant pigs. *Applied Animal Behaviour Science* 20, 287–296.

Barnett, J.L., Hemsworth, P.H., Newman, E.A., McCallum, T.H. and Winfield, C.G. (1989) The effect of design of tether and stall housing on the behavioural and physiological responses related to the welfare of pregnant pigs. *Applied Animal Behaviour Science* 24, 1–12.

Barnett, J.L., Hemsworth, P.H., Cronin, G.M., Newman, E.A. and McCallum, T.H. (1991) Effects of design of individual cage-stalls on the behavioural and physiological responses related to the welfare of pregnant pigs. *Applied Animal Behaviour Science* 32, 23–33.

Barnett, J.L., Hemsworth, P.H. and Newman, E.A. (1992) Fear of humans and its relationships with productivity in laying hens at commercial farms. *British Poultry Science* 33, 699–710.

Barnett, J.L., Hemsworth, P.H. and Jones, R.B. (1993) Behavioural responses of commercial farmed laying hens to humans: evidence of stimulus generalization. *Applied Animal Behaviour Science* 37, 139–146.

Barnett, J.L., Hemsworth, P.H., Hennessy, D.P., McCallum, T.M. and Newman, E.A. (1994) The effects of modifying the amount of human contact on the behavioural, physiological and production responses of laying hens. *Applied Animal Behaviour Science* 41, 87–100.

Barrick, M.P. and Mount, M. (1991) The big five personality dimensions and job performance: a meta-analysis. *Personnel Psychology* 44, 1–26.

Barrick, M.R., Mount, M.K. and Strauss, P.S. (1993) Conscientiousness and performance of sales representatives: test of the mediating effects of goal setting. *Journal of Applied Psychology* 78(5), 715–722.

Barton Gade, P., Blaabjerg, L. and Christensen, L. (1992) New lairage system for slaughter pigs – effects on behaviour and quality characteristics. In: *Proceedings 38th International Congress of Meat Science and Technology, Clermont-Ferrand, France*, Vol. II. Danish Meat Institute, Roskilde, pp. 161–164.

Bartov, I., Jensen, L.S. and Veltman, J.R. (1980) Effect of corticosterone and prolactin on fattening in broiler chicks. *Poultry Science* 59, 1328–1334.

Beilharz, R.G. (1982) Genetic adaptation in relation to animal welfare. *International Journal of the Study of Animal Problems* 3, 117–124.

Beilharz, R.G. and Zeeb, K. (1981) Applied ethology and animal welfare. *Applied Animal Ethology* 7, 3–10.

Bellamy, D. and Leonard, R.A. (1965) Effect of cortisol on the growth of chicks. *General Comparative Endocrinology* 5, 402–410.

Bentler, P.M. and Speckart, G. (1979) Models of attitude–behaviour relations. *Psychological Review* 86(5), 452–464.

Beveridge, L.M. (1996) Studies on the influence of human characteristics and training on stockperson work performance and farm animal behaviour. PhD thesis, University of Aberdeen, UK.

Beynon, N.M. (1991) Analysis of stockmanship. *Pig Veterinary Journal* 26, 67–77.

Blake, R. and Dennis, W. (1943) Development of stereotypes concerning the Negro. *Journal of Abnormal Social Psychology* 38, 525–531.

Blumberg, M. and Pringle, C.D. (1982) The missing opportunity in organisational research: some implications for a theory of work performance. *Academy of Management Review* 7(4), 560–569.

Boissy, A. and Bouissou, M.F. (1988) Effects of early handling on heifers' subsequent reactivity to humans and to unfamiliar situations. *Applied Animal Behaviour Science* 20, 259–273.

Boivin, X., Le Neindre, P. and Chupin, J.M. (1992) Establishment of cattle–human relationships. *Applied Animal Behaviour Science* 32, 325–335.

Borofsky, G.L. and Smith, M. (1993) Reductions in turnover, accidents, and absenteeism: the contribution of a pre-employment screening inventory. *Journal of Clinical Psychology* 49(1), 109–116.

Bouissou, M.F. and Vandenheede, M. (1995) Fear reactions of domestic sheep confronted with either a human or a human-like model. *Behavioural Processes* 43(1), 81–92.

Bradley, A.J., McDonald, I.R. and Lee, A.K. (1980) Stress and mortality in a small marsupial (*Antechinus stuartii*, Macleay). *General Comparative Endocrinology* 40, 188–200.

Bredbacka, P. (1988) Relationships between fear, welfare and productive traits in caged White Leghorn hens. In: Unshelm, J., van Putten, G., Zeeb, K. and Ekesbo, I. (eds) *Proceedings of the International Congress on Applied Ethology in Farm Animals*. Research Institute of Animal Production, Prague, Czech Republic, and Institute of Animal Biochemistry and Genetics, Slovak Academy of Sciences, Slovakia, pp. 74–89.

Brehm, J. and Cohen, A.R. (1962) *Explorations in Cognitive Dissonance*. Wiley, New York.

Breuer, K., Hemsworth, P.H. and Coleman, G.J. (1997) The influence of handling on the behaviour and productivity of lactating heifers. In: *Proceedings of the 31st International Congress of the International Society for Applied Ethology, Prague, Czech Republic*. Research Institute of Animal Production, Prague, Czech Republic, and Institute of Animal Biochemistry and Genetics, Slovak Academy of Sciences, Slovakia.

Broom, D.M. (1986) Indicators of poor welfare. *British Veterinary Journal* 142, 524–526.

Broom, D.M and Johnson, K.G. (1993) *Stress and Animal Welfare*. Chapman and Hall, London.

Buckland, R.B., Goldrosen, A. and Bernon, D.E. (1974) Effect of blood sampling by cardiac puncture on subsequent body weight of broilers and S.C. White Leghorn replacement pullets. *Poultry Science* 53, 1256–1258.

Budd, R.J., North, D. and Spencer, C. (1984) Understanding seat-belt use: a test of Bentler and Speckart's extension of the 'theory of reasoned action'. *European Journal of Social Psychology* 14(1), 69–78.

Burrow, H.M. and Dillion, R.D. (1997) The relationship between temperament, liveweights and commercial carcass traits of *Bos indicus* crossbreds. *Australian Journal of Experimental Agriculture* (in press).

Burrow, H.M., Seifert, G.W. and Corbert, N.J. (1988) A new technique for measuring temperament in cattle. *Proceedings of the Australian Society of Animal Production* 17, 154–157.

Cain, A.O. (1991) Pets and the family. *Holistic Nursing Practice* 5(2), 58–63.

Caine, N.G. (1992) Humans as predators: observational studies and the risk of pseudohabituation. In: Davis, H. and Balfour, A.D. (eds) *The Inevitable Bond – Examining Scientist–Animal Interactions*. Cambridge University Press, Cambridge, pp. 357–364.

Campbell, D.T. (1963) Social attitudes and other acquired behavioural dispositions. In: Koch, S. (ed.) *Psychology: a Study of a Science*. McGraw-Hill, New York, pp. 94–172.

Campion, M.A. (1991) Meaning and measurement of turnover: comparison of alternative measures and recommendations for research. *Journal of Applied Psychology* 76(2), 199–212.

Cannon, W.B. (1914) The emergency function of the adrenal medulla in pain and the major emotions. *American Journal of Physiology* 33, 356–372.

Cheal, P.D., Lee, A.K. and Barnett, J.L. (1976) Changes in the haematology of *Antechinus stuartii* (Marsupialia), and their association with male mortality. *Australian Journal of Zoology* 24, 299–311.

Choplan, B.E., McCain, M.L., Carbonell, J.L. and Hagen, R. (1985) Empathy: review of available measures. *Journal of Personality and Social Psychology* 48(3), 635–653.

Chusmir, L.H. (1982) Job commitment and the organisational woman. *Academy of Management Review* 7, 595–602.

Clarke, I.J., Hemsworth, P.H., Barnett, J.L. and Tilbrook, A.J. (1992) Stress and reproduction in farm animals. In: Sheppard, K.E., Boublik, J.H. and Funder, J.W. (eds) *Stress and Reproduction*. Serono Symposium Publications, Vol. 86, Raven Press, New York, pp. 239–251.

Cleary, G.V. (1990) Personnel management and staff training. In: Gardner, J.A.A., Dunkin, A.C. and Lloyd, L.C. (eds) *Personnel Management and Record Keeping: Pig Production in Australia*. Australian Pig Research Council, Butterworths, Canberra.

Coleman, G.J., Hemsworth, P.H., Hay, M. and Cox, M. (1998) Predicting stockperson behaviour towards pigs from attitudinal and job-related variables and empathy. *Applied Animal Behaviour Science* (in press).

Collins, J.W. and Siegel, P.B. (1987) Human handling, flock size and responses to an *E. coli* challenge in young chickens. *Applied Animal Behavior Science* 19, 183–188.

Cooper, J.J. and Nicol, C.J. (1991) Stereotypic behaviour affects environmental preference in bank voles, *Clethrionomys glareolus*. *Animal Behaviour* 41, 971–977.

Cox, R.P. (1993) The human–animal bond as a correlate of family functioning. *Clinical Nursing Research* 2(2), 224–231.

Cransberg, P.H. (1996) The relationship between human factors and the productivity and welfare of commercial broiler chickens. MBehavSci thesis, La Trobe University, Victoria, Australia.

Cransberg, P.H. and Hemsworth, P.H. (1995) Human factors affecting the behavioural response and productivity of commercial broiler chickens. *Queensland Poultry Science Symposium* 4, 1–8.

Creel, S.R. and Albright, J.L. (1988) The effect of neonatal social isolation of the behaviour and endocrine function of Holstein calves. *Applied Animal Behaviour Science* 21, 293–306.

Cronin, G.M., Wiepkema, P.R. and van Ree, J.M. (1986) Endorphins implicated in stereotypies of tethered sows. *Experiential* 42, 198–199.

Davis, H. and Balfour, A.D. (1992) The inevitable bond. In: Davis, H. and Balfour, A.D. (eds) *The Inevitable Bond – Examining Scientist–Animal Interactions.* Cambridge University Press, Cambridge, pp. 27–43.

Dawkins, M. (1983) Battery hens name their price: consumer demand theory and the measurement of animal needs. *Animal Behaviour* 31, 1195–1205.

Dawkins, M.S. (1990) From an animal's point of view: motivation, fitness and animal welfare. *Behavioural and Brain Science* 13, 1–61.

de Passille, A.M., Rushen, J., Ladewig, J. and Petherick C. (1996) Dairy calves' discrimination of people based on previous handling. *Journal of Animal Science* 74, 969–974.

Dewsbury, D.A. (1992) Studies of rodent–human interactions in animal psychology. In: Davis, H. and Balfour, A.D. (eds) *The Inevitable Bond – Examining Scientist–Animal Interactions.* Cambridge University Press, Cambridge, pp. 27–43.

Duncan, I.J.H. (1978) The interpretation of preference tests in animal behaviour. *Applied Animal Ethology* 4, 197–200.

Duncan, I.J.H. and Petherwick, J.C. (1991) The implications of cognitive processes for animal welfare. *Journal of Animal Science* 69, 5017–5022.

Duncan, I.J.H., Slee, G.S., Kettlewell, P., Berry, P. and Carlisle, A.J. (1986) Comparison of the stressfulness of harvesting broiler chickens by machine and by hand. *British Poultry Science* 27, 109–114.

Eagly, A.H. and Chaiken, S. (1993) *The Psychology of Attitudes.* Harcourt Brace Jovanovitch, Orlando, Florida.

Echabe, A.E., Rovir, D.P. and Garate, J.V. (1988) Testing Ajzen and Fishbein's attitude model: the prediction on voting. *European Journal of Social Psychology* 18(2), 181–189.

Edney, A.T.B. (1992) Companion animals and human health. *Veterinary Record* 130(14), 285–288.

English, P., Burgess, G., Segundo, R. and Dunne, J. (1992) *Stockmanship: Improving the Care of the Pig and Other Livestock.* Farming Press Books, Ipswich, UK.

Estep, D.Q. and Hetts, S. (1992) Interactions, relationships, and bonds: the conceptual basis for scientist–animal relations. In: Davis, H. and Balfour, A.D. (eds) *The Inevitable Bond – Examining Scientist–Animal Interactions.* Cambridge University Press, Cambridge, pp. 6–26.

Ewbank, R. (1978) Stereotypies in clinical veterinary practice. In: *1st World Congress on Ethology Applied to Zootechnics.* Industrias Graficas España, Madrid, pp. 499–502.

Fell, L.R. and Shutt, D.A. (1986) Adrenocortical response of calves to transport stress as measured by salivary cortisol. *Canadian Journal of Animal Science* 66, 637–641.

Festinger, L. (1957) *A Theory of Cognitive Dissonance*. Stanford University Press, Stanford, California.

Festinger, L. and Carlsmith, J.M. (1959) Cognitive consequences of forced compliance. *Journal of Abnormal and Social Psychology* 58, 203–210.

Fishbein, M. and Ajzen, I. (1974) Attitudes towards objects as predictors of single and multiple behavioral criteria. *Psychological Review* 81(1), 59–74.

Fordyce, G., Wythes, J.R., Shorthose, W.R., Underwood, D.W. and Shepherd, R.K. (1988) Cattle temperaments in extensive beef herds in northern Queensland. 2. Effect of temperament on carcass and meat quality. *Australian Journal of Experimental Agriculture* 28, 689–693.

Fraser, A.F. and Broom, D.M. (1990) *Farm Animal Behaviour and Welfare*. CAB International, Wallingford, UK.

Freeman, B.M. and Manning, A.C.C. (1979) Stressor effects of handling on the immature fowl. *Research in Veterinary Science* 26, 223–226.

Galef, B.G., Jr (1970) Aggression and timidity: responses to novelty in feral Norway rats. *Journal of Comparative and Physiological Psychology* 70(3), 370–381.

Gonyou, H.W. (1993) Behavioural principles of animal handling and transport. In: Grandin, T. (ed.) *Livestock Handling and Transport*. CAB International, Wallingford, UK, pp. 11–20.

Gonyou, H.W., Hemsworth, P.H. and Barnett, J.L. (1986) Effects of frequent interactions with humans on growing pigs. *Applied Animal Behaviour Science* 16, 269–278.

Grandin, T. (1980) Livestock behavior as related to handling facilities design. *International Journal for the Study of Animal Problems* 1, 33–52.

Grandin, T. (1993) Behavioural principles of cattle handling under extensive conditions. In: Grandin, T. (ed.) *Livestock Handling and Transport*. CAB International, Wallingford, UK, pp. 43–57.

Grandin, T., Curtis, S.E. and Taylor, I.A. (1987) Toys, mingling and driving reduce excitability in pigs. *Journal of Animal Science* 65 (Suppl. 1), 230 (abstract).

Gray, J.A. (1987) *The Psychology of Fear and Stress*, 2nd edn. Cambridge University Press, Cambridge.

Gregory, N.G., Wilkins, L.J., Alvey, D.M. and Tucker, S.A. (1993) Effect of catching method and lighting intensity on the prevalence of broken bones and on the ease of handling of end of lay hens. *Veterinary Record* 132, 127–129.

Griffin, D.R. (1992) *Animal Minds*. University of Chicago Press, London.

Gross, W.B. and Siegel, P.B. (1979) Adaptation of chickens to their handlers and experimental results. *Avian Diseases* 23(3), 708–714.

Gross, W.B. and Siegel, P.B. (1980) Effects of early environmental stresses on chicken body weight, antibody response to RBC antigens, feed efficiency and response to fasting. *Avian Diseases* 24, 549–579.

Gross, W.B. and Siegel, P.B. (1982) Influences of sequences of environmental factors on the responses of chickens to fasting and to *Staphylococcus aureus* infection. *American Journal of Veterinary Research* 43, 137–139.

Hale, E.B. (1969) Domestication and the evolution of behaviour. In: Hafez, E.S.E. (ed.) *The Behavior of Domestic Animals*, 2nd edn. Williams and Wilkins, Baltimore.

Hargreaves, A.L. and Hutson, G.D. (1990) The effect of gentling on heart rate, flight distance and aversion of sheep to a handling procedure. *Applied Animal Behaviour Science* 26, 243–252.

Harrison, R. (1964) *Animal Machines*. Stuart, London.

Hearnshaw, H., Barlow, R. and Want, G. (1979) Development of a 'temperament' or 'handling difficulty' score for cattle. In: *Proceeding of the Inaugural Conference on Australian Animal Breed Genetics,* Vol. 1. Australian Association of Animal Breeding and Genetics, University of New England, Armidale, Australia, pp. 164–166.

Hediger, H. (1964) The animal's relationship with man. In: *Wild Animals in Captivity*. Dover Publications, New York.

Hemsworth, P.H. and Barnett, J.L. (1987) Human–animal interactions. In: Price, E.O. (ed.) *The Veterinary Clinics of North America, Food Animal Practice*, Vol. 3. W.B. Saunders, Philadelphia, pp. 339–356.

Hemsworth, P.H. and Barnett, J.L. (1989) Relationships between fear of humans, productivity and cage position of laying hens. *British Poultry Science* 30, 505–518.

Hemsworth, P.H. and Barnett, J.L. (1991) The effects of aversively handling pigs either individually or in groups on their behaviour, growth and corticosteroids. *Applied Animal Behaviour Science* 30, 61–72.

Hemsworth P.H. and Gonyou, H.W. (1997) Human contact. In: Appleby, M.C. and Hughes, B.O. (eds) *Animal Welfare*. CAB International, Wallingford, UK, pp. 205–305.

Hemsworth, P.H., Barnett, J.L. and Hansen, C. (1981a) The influence of handling by humans on the behaviour, growth and corticosteroids in the juvenile female pig. *Hormones and Behavior* 15, 396–403.

Hemsworth, P.H., Brand, A. and Willems, P. (1981b) The behavioural response of sows to the presence of human beings and its relation to productivity. *Livestock Production Science* 8, 67–74.

Hemsworth, P.H., Gonyou, H.W. and Dzuik, P.J. (1986a) Human communication with pigs: the behavioural response of pigs to specific human signals. *Applied Animal Behaviour Science* 15, 45–54.

Hemsworth, P.H., Barnett, J.L. and Hansen, C. (1986b) The influence of handling by humans on the behaviour, reproduction and corticosteroids of male and female pigs. *Applied Animal Behaviour Science* 15, 303–314.

Hemsworth, P.H., Barnett, J.L., Hansen, C. and Winfield, C.G. (1986c) Effects of social environment on welfare status and sexual behaviour of female pigs. II. Effects of space allowance. *Applied Animal Behaviour Science* 16, 259–267.

Hemsworth, P.H., Barnett, J.L. and Hansen, C. (1987) The influence of inconsistent handling on the behaviour, growth and corticosteroids of young pigs. *Applied Animal Behaviour Science* 17, 245-252.

Hemsworth, P.H., Barnett, J.L., Coleman, G.J. and Hansen, C. (1989) A study of the relationships between the attitudinal and behavioural profiles of stockpeople and the level of fear of humans and the reproductive performance of commercial pigs. *Applied Animal Behaviour Science* 23, 301–314.

Hemsworth, P.H., Barnett, J.L. and Coleman, G.J. (1993) The human–animal relationship in agriculture and its consequences for the animal. *Animal Welfare* 2, 33–51.

Hemsworth, P.H., Coleman, G.J. and Barnett, J.L. (1994a) Improving the attitude and behaviour of stockpeople towards pigs and the consequences on the behaviour and reproductive performance of commercial pigs. *Applied Animal Behaviour Science* 39, 349–362.

Hemsworth, P.H., Coleman, G.J., Cox, M. and Barnett, J.L. (1994b) Stimulus generalisation: the inability of pigs to discriminate between humans on the basis of their previous handling experience. *Applied Animal Behaviour Science* 40, 129–142.

Hemsworth, P.H., Coleman, G.J., Barnett, J.L. and Jones, R.B. (1994c) Behavioural responses of humans and the productivity of commercial broiler chickens. *Applied Animal Behaviour Science* 41, 101–114.

Hemsworth, P.H., Barnett, J.L., Breuer, K., Coleman, G.J. and Matthews, L.R. (1995) *An Investigation of the Relationships between Handling and Human Contact and the Milking Behaviour, Productivity and Welfare of Commercial Dairy Cows.* Research Report on Dairy Research and Development Council Project, Attwood, Australia.

Hemsworth, P.H., Coleman, G.J., Cransberg, P.H. and Barnett, J.L. (1996a) *Human Factors and the Productivity and Welfare of Commercial Broiler Chickens.* Research Report on Chicken Meat Research and Development Council Project, Attwood, Australia.

Hemsworth, P.H., Barnett, J.L. and Campbell, R.G. (1996b) A study of the relative aversiveness of a new daily injection procedure for pigs. *Applied Animal Behaviour Science* 49, 389–401.

Hemsworth, P.H., Price, E.O. and Bogward, R. (1996c) Behavioural responses of domestic pigs and cattle to humans and novel stimuli. *Applied Animal Behavioural Science* 50, 43–56.

Hemsworth, P.H., Verge, J. and Coleman, G.J. (1996d) Conditioned approach avoidance responses to humans: the ability of pigs to associate feeding and aversive social experiences in the presence of humans with humans. *Applied Animal Behaviour Science* 50, 71–82.

Hinde, R.A. (1970) *Behaviour: a Synthesis of Ethology and Comparative Psychology.* McGraw-Hill, Kogakusha, Japan.

Horowitz, E. and Horowitz, R. (1938) Development of social attitudes in children. *Sociometry* 1, 301–338.

Hovland, C.I., Janis, I.L. and Kelley, H.H. (1953) *Communication and Persuasion: Psychological Studies of Opinion Change.* Yale University Press, New Haven, Connecticut.

Hughes, B.O. (1975) Spatial preference in the domestic fowl. *British Veterinary Journal* 131, 560–564.

Hughes, B.O. and Black, A.J. (1976) The influence of handling on egg production, egg shell quality and avoidance behaviour of hens. *British Poultry Science* 17, 135–144.

Hughes, B.O. and Duncan, I.J.H. (1988) Behavioural needs: can they be explained in terms of motivational models? *Applied Animal Behaviour Science* 20, 352–355.

Hutson, G.D. (1982) 'Flight distance' in merino sheep. *Animal Production* 35, 231–235.

Hutson, G.D. (1985) The influence of barley food rewards on sheep movement through a handling system. *Applied Animal Behaviour Science* 14, 263–273.

Hutson, G.D. (1992) A comparison of operant responding by farrowing sows for food and nest-building material. *Applied Animal Behaviour Science* 34, 221–230.

Ivanov, Y.K. and Patrushev, V.D. (1976) The impact of labor conditions on work satisfaction of agricultural workers. *Sotisiologicheskie-Issledovaniya* 3 (July–September), 60–70.

Janis, I.L. and Hovland, C.I. (1959) An overview of persuasibility research. In: Hovland, C.I. and Hanis, I.L. (eds) *Personality and Persuasibility*. Yale University Press, New Haven, Connecticut, pp. 1–28.

Jewell, L.N. and Siegall, M. (1990) *Contemporary Industrial/Organizational Psychology*, 2nd edn. West, St Paul.

Johns, G. (1987) The great escape: the whys and wherefores of absenteeism add up to a $30 billion question for American business. *Psychology Today* 21, 30–33.

Jones, R.B. (1987) Fear and fear responses: a hypothetical consideration. *Medical Science Research* 15, 1287–1290.

Jones, R.B. (1993) Reduction of the domestic chick's fear of humans by regular handling and related treatments. *Animal Behaviour* 46, 991–998.

Jones, R.B. (1994) Regular handling and the domestic chick's fear of human beings: generalisation of response. *Applied Animal Behaviour Science* 42, 129–143.

Jones, R.B. and Faure, J.M. (1981) The effects of regular handling on fear responses in the domestic chick. *Behavioural Processes* 6, 135–143.

Jones, R.B. and Hughes, B.O. (1981) Effects of regular handling on growth in male and female chicks of broiler and layer strains. *British Poultry Science* 22, 461–465.

Jones, R.B. and Waddington, D. (1992) Modification of fear in domestic chicks, *Gallus gallus domesticus* via regular handling and early environmental enrichment. *Animal Behaviour* 43, 1021–1033.

Jones, R.B. and Waddington, D. (1993) Attenuation of the domestics chick's fear of human beings via regular handling: in search of a sensitive period. *Applied Animal Behaviour Science* 36, 185–195.

Jones, R.B., Duncan, I.J.H. and Hughes, B.O. (1981) The assessment of fear in domestic hens exposed to a looming human stimulus. *Behavioural Processes* 6, 121–133.

Jones, R.B., Mills, A.D. and Faure, J.M. (1991) Genetic and experimental manipulation of fear-related behaviour in Japanese quail chicks (*Coturnix coturnix japonica*). *Journal of Comparative Psychology* 105, 15–24.

Kanekar, S. (1976) Observational learning of attitudes: a behavioural analysis. *European Journal of Social Psychology* 6(1), 5–24.

Katz, D. (1960) The functional approach to the study of attitude. *Public Opinion Quarterly* 24, 163–204.

Kelley, K.W. (1985) Immunological consequences of changing environmental stimuli. In: Moberg, G.P. (ed.) *Animal Stress*. American Physiological Society, Baltimore, USA.

Kendall, P.C. and Hollon, S.D. (1979) *Cognitive–Behavioural Interventions*. Academic Press, London.

Kent, J.E. and Ewbank, R. (1983) The effect of road transportation on the blood constituents and behaviour of calves. I. Six months old. *British Veterinary Journal* 139, 228–234.

Kent, J.E. and Ewbank, R. (1986a) The effect of road transportation on the blood constituents and behaviour of calves. II. One to three weeks old. *British Veterinary Journal* 142, 131–140.

Kent, J.E. and Ewbank, R. (1986b) The effect of road transportation on the blood constituents and behaviour of calves. III. Three months old. *British Veterinary Journal* 142, 326–335.

Klasing, K.C. (1985) Influence of stress on protein metabolism In: Moberg, G.P. (ed.) *Animal Stress*. American Physiological Society, Baltimore, USA.

Kondos, A.C. (1983) Human resources. In: *Proceedings of the 2nd National Pig Production and Marketing Review Conference, Hobart*. Standing Committee on Agriculture, Department of Primary Industries and Energy, Canberra, Australian Capital Territory, Australia.

La Piere, R.T. (1934) Attitudes vs. actions. *Social Forces* 13, 230–237.

Lawrence, A.B. and Illius, A.W. (1997) Measuring preferences and the problems of identifying proximate needs. In: Forbes, J.M., Lawrence, T.L.J., Rodway, R.G. and Varley, M.A. (eds) *Animal Choices*. British Society of Animal Science, Occasional Publication no. 20, BSAS, Edinburgh, pp. 19–26.

Lee, A.K., Bradley, A.J. and Braithwaite, R.W. (1977) Corticosteroid levels and male mortality in *Antechinus stuartii*. In: Stonehouse, B. and Gilmore, D. (eds) *The Biology of Marsupials*. MacMillan Press, London, pp. 209–220.

Leslie, B.E., Meek, A.H., Kawash, G.F. and McKeown, D.B. (1994) An epidemiological investigation of pet ownership in Ontario. *Canadian Veterinary Journal* 35, 218–222.

Levine, S. (1985) A definition of stress. In: Moberg, G.P. (ed.) *Animal Stress*. American Physiological Society, Bethesda, Maryland, pp. 51–69.

Lloyd, D.H. (1974) Effective staff management. In: Freedman, B.M. and Boorman, K.N. (eds) *Economic Factors Affecting Egg Production*. British Poultry Science, Edinburgh, pp. 221–251.

Lorenz, K.Z. (1970) Companions as factors in the bird's environment. In: Lorenz, K.Z. (ed.) *Studies on Animal and Human Behaviour*. Harvard University Press, Cambridge, Massachusetts.

Lynne, G.D. and Rola, L.R. (1987) Improving attitude–behaviour prediction models with economic variables: farmer actions towards soil conservation. *Journal of Social Psychology* 128(1), 19–28.

Lyons, D.M. (1989) Individual differences in temperament of dairy goats and the inhibition of milk ejection. *Applied Animal Behaviour Science* 22, 269–282.

Lyons, D.M., Price, E.D. and Moberg, G.P. (1988) Individual differences in temperament of domestic dairy goats: constancy and change. *Animal Behaviour* 36, 1323–1333.

McFarland, D. (1981) *The Oxford Companion to Animal Behaviour*. Oxford University Press, Oxford.

McGuire, W.J. (1964). Inducing resistance to persuasion: some contemporary approaches. In: Berkowitz, L. (ed.) *Advances in Experimental Social Psychology*, Vol. 1. Academic Press, New York, pp. 192–227.

Maio, G.R., Esses, V.M. and Bell, D.W. (1994) The formation of attitudes toward new immigrant groups. *Journal of Applied Social* Psychology 24(19), 1762–1776.

Maslow, A.H. (1970) *Motivation and Personality*, 2nd edn. Harper and Row, New York.

Mason, G. and Mendl, M. (1993) Why is there no simple way of measuring animal welfare? *Animal Welfare* 2, 301–319.

Mason, G.J. (1991) Stereotypies: a critical review. *Animal Behaviour* 41, 1015–1037.

Mason, J.W. (1968a) The scope of psychoendocrine research. *Psychomatic Medicine* 30, 565–575.

Mason, J.W. (1968b) A review of psychoendocrine research on the pituitary–adrenal cortical system. *Psychosomatic Medicine* 30, 576–607.

Mason, J.W. (1971) A re-evaluation of the concept of 'non-specificity' in stress theory. *Journal of Psychiatric Research* 8, 323–333.

Mateo, J.M., Estep, D.Q. and McCann, J.S. (1991) Effects of differential handling on the behaviour of domestic ewes (*Ovis aries*). *Applied Animal Behaviour Science* 32, 45–54.

Matthews, L.R. (1993) Deer handling and transport. In: Grandin, T. (ed.) *Livestock Handling and Transport*. CAB International, Wallingford, pp. 253–272.

Miller, D. (1995) Evaluating stockmanship capabilities in employees. In: *Professional Swine Management Certification Series, Swine Breeding, Herd Management, Module 3, Stockmanship*, Part H. Iowa Pork Industry Center, Ames, Iowa, pp. 1–4.

Mills, A.D. and Faure, J.-M. (1990) Panic and hysteria in domestic fowl: a review. In: Zayan, R. and Dantzer, R. (eds) *Social Stress in Domestic Animals*. Kluwer Academic Publishers, Dordrecht, pp. 248-272.

Ministry of Agriculture, Fisheries and Food (1983) *British Codes of Recommendations for the Welfare of Livestock*. HMSO, London.

Moberg, G.P. (1985) Influence of stress on reproduction: measure of well-being. In: Moberg, G.P. (ed.) *Animal Stress*. American Physiological Society, Bethesda, Maryland, pp. 245–267.

Munksgaard, L., de Passille, A.M., Rushen, J., Thodberg, K. and Jensen, M.B. (1995) The ability of dairy cows to distinguish between people. In: *Proceedings of the 29th International Congress of the International Society of Applied Ethology, 3–5 August, Exeter*. University Federation for Animal Welfare, Potters Bar, UK, pp. 19–20 (abstract).

Murphey, R.M., Moura Duarte, F.A. and Torres Penendo, M.C. (1981) Responses of cattle to humans in open spaces: breed comparisons and approach–avoidance relationships. *Behaviour Genetics* 2(1), 37–47.

Murphy, L.B. (1976) A study of the behavioural expression of fear and exploration in two stocks of domestic fowl. PhD dissertation, Edinburgh University, UK.

Murphy, L.B. (1978) The practical problems of recognizing and measuring fear and exploration behaviour in the domestic fowl. *Animal Behaviour* 26, 422–431.

Murphy, L.B. and Duncan, L.J.H. (1977) Attempts to modify the responses of domestic fowl towards human beings. I. The association of human contact with a food reward. *Applied Animal Ethology* 3, 321–334.

Murphy, L.B. and Duncan, L.J.H. (1978) Attempts to modify the responses of domestic fowl towards human beings. II. The effect of early experience. *Applied Animal Ethology* 4, 5–12.

Nichol, C.J. (1995) Cognition: a thoughtful approach to behaviour. In: *Proceedings of the 29th International Congress of the International Society of Applied Ethology, 3–5 August, Exeter*. University Federation for Animal Welfare, Potters Bar, UK, pp. 3–9.

Olweus, D. (1979) Stability of aggressive reaction patterns in males: a review. *Psychology Bulletin* 86, 852–875.

Paterson, A.M. and Pearce, G.P. (1989) Boar-induced puberty in gilts handled pleasantly or unpleasantly during rearing. *Applied Animal Behaviour Science* 22, 225–233.

Paterson, A.M. and Pearce, G.P. (1992) Growth, response to humans and corticosteroids in male pigs housed individually and subjected to pleasant, unpleasant or minimal handling during rearing. *Applied Animal Behaviour Science* 34, 315–328.

Pavlov, I.P. (1960) *Conditioned Reflexes: an Investigation of the Physiological Activity of the Cerebal Cortex.* Dover, New York.

Pearce, G.P., Paterson, A.M. and Pearce, A.N. (1989) The influence of pleasant and unpleasant handling and the provision of toys on the growth and behaviour of male pigs. *Applied Animal Behaviour Science* 23, 27–37.

Pedersen, V. (1993) Effects of difference post-weaning handling procedures on the later behaviour of silver foxes. *Applied Animal Behaviour Science* 37, 239–250.

Pedersen, V. and Jeppesen, L.L. (1990) Effects of early handling on better behaviour and stress responses in the silver fox (*Vulpes vulpes*). *Applied Animal Behaviour Science* 26, 383–393.

Pedersen, V., Barnett, J.L., Hemsworth, P.H., Newman, E.A. and Schirmer B. (1997) The effects of handling on behavioural and physiological responses to housing in tether-stalls in pregnant pigs. *Animal Welfare* (in press).

Phillips, J.C. and Benson, J.E. (1983) Some aspects of job satisfaction in the Soviet Union. *Personnel Psychology* 36(3), 633–645.

Pianka, E.R. (1974) *Evolutionary Ecology*. Harper and Row, New York.

Price, E.O. (1984) Behavioral aspects of animal domestication. *Quarterly Review of Biology* 59, 1–32.

Purcell, D., Arave, C.W. and Walters, J.L. (1988) Relationship of three measures of behaviour to milk production. *Applied Animal Behaviour Science* 21, 307–313.

Ravel, A., D'Allaire, S.D., Bigras-Poulin, M. and Ward, R. (1996) Personality traits of stockpeople working in farrowing units on two types of farms in Quebec. In: *Proceedings of 14th Congress of the International Pig Veterinary Society, 7–10 July, Bologna, Italy.* Faculty of Veterinary Medicine, University of Bologna, Bologna, p. 514.

Ravel, A., D'Allaire, S., Bigras-Poulin, M. and Ward, R. (1997) Influence of management, housing and personality of the stockperson on preweaning performances on independent and integrated swine farms in Quebec. *Preventive Veterinary Medicine* (in press).

Reber, A.S. (1988) *Dictionary of Psychology*. Penguin Books, London.

Reichmann, K.G., Barram, K.M., Brock, I.J. and Standfast, N.F. (1978) Effects of regular handling and blood sampling by wing vein puncture on the performance of broilers and pullets. *British Poultry Science* 19, 97–99.

Rivier, C. and Rivest, S. (1991) Effect of stress on the activity of the hypothalamic–pituitary–gonadal axis: peripheral and central mechanisms. *Biology of Reproduction* 45, 523–532.

Rokeach, M. (1960) *The Open and Closed Mind: Investigations into the Nature of Belief Systems and Personality Systems.* Basic Books, New York.

Rural Training Council of Australia (1990) *Australian National Pig Industry Training Program*. Rural Training Council of Australia.

Rusbult, C.E, Farrell, D., Rogers, G. and Mainous, A.G., III (1988) Impact of exchange variables on exit, voice, loyalty, and neglect: an integrative model of responses to declining job satisfaction. *Academy of Management Journal* 31, 599–627.

Rushen, J. (1993) The coping hypothesis of stereotypic behaviour. *Animal Behaviour* 45, 613–615.

Rushen, J., Munksgaard L., de Passille, A.M.B., Jensen, M.B. and Thodberg K. (1995) Location of handling and dairy cows' ability to discrimate between gentle and aversive handlers. In: *Proceedings of the 29th International Congress of the International Society for Applied Ethology, Exeter, UK.* University Federation for Animal Welfare, Potter's Bar, UK, p. 219 (abstract).

Rushen, J., de Passille, A.M. and Munksgaard, L. (1997) Dairy cows' fear of people reduces milk yield and affects behaviour at milking. In: *Proceedings of the 31st International Congress of the International Society for Applied Ethology, Prague, Czech Republic.* Research Institute of Animal Production, Prague, and Institute of Animal Biochemistry and Genetics, Slovak Academy of Sciences, Slovakia, p. 215 (abstract).

Saadoun, A., Simon, J.R. and Leclercq, B. (1987) Effect of exogenous corticosterone in genetically fat and lean chickens. *British Poultry Science* 28, 519–528.

Schaefer, T. (1968) Some methodological implication of the research on 'early handling' in the rat. In: Newton, G. and Levine, S. (eds) *Early Experience and Behaviour: The Psychobiology of Development.* Charles C. Thomas, Springfield, Illinois.

Scott, J.P. (1992) The phenomenon of attachment in human–non-human relationships. In: Davis, H. and Balfour, A.D. (eds) *The Inevitable Bond – Examining Scientist–Animal Interactions.* Cambridge University Press, Cambridge, pp. 72–92.

Seabrook, M.F. (1972a) A study to determine the influence of the herdsman's personality on milk yield. *Journal of Agriculture Labour Science* 1, 45–59.

Seabrook, M.F. (1972b) A study of the influence of the cowman's personality and job satisfaction on milk yield of dairy cows. In: *Joint Conference of the British Society for Agriculture Labour Science and the Ergonomics Research Society, September 1972, National College of Agricultural Engineering, UK.*

Seabrook, M.F. (1994) The effect of production systems on the behaviour and attitudes of stockpersons. In: *4th Zodiac Symposium.* Publication No. 67, EAPP, Wageningen, the Netherlands, pp. 252–258.

Seabrook, M.F. and Bartle, N.C. (1992) The practical implications of animals' responses to man. In: *Winter Meeting, British Society of Animal Production, 23–25 March 1992, Scarborough, UK.* Paper no. 34.

Segundo, R.C. (1989) A study of stockpeople and managers in the pig industry with special emphasis on the factors affecting their job satisfaction. MSc thesis, University of Aberdeen, Aberdeen, UK.

Selye, H. (1946) The general adaptation syndrome and the diseases of adaptation. *Journal of Clinical Endocrinology* 6, 117–230.

Selye, H. (1976) *Stress in Health and Disease.* Butterworths, Boston.

Siegel, H.S. and van Kampen, M. (1984) Energy relationships in growing chickens given daily injections of corticosterone. *British Poultry Science* 25, 477–485.

Singer, P. (1975) *Animal Liberation.* Avon Books, New York.

Singh, S. (1983) Effect of motivation, values, cognitive factors and child rearing among Punjab farmers. *Journal of Social Psychology* 120(2), 273–278.

Skinner, B.F. (1969) *Contingencies of Reinforcement: a Theoretical Analysis.* Prentice Hall, Englewood Cliffs, New Jersey.

Stagner, R. (1961) *Psychology of Personality.* McGraw-Hill, New York.

Syme, L.A., Durham, I.H. and Elphick, G.R. (1981) Microprocessor control of sheep movement. In: *Proceedings of the 2nd Conference on Wool Harvesting Research and Development*. Australian Wool Corporation, Sydney, pp. 237-245.

Tanida, H., Miura, A., Tanaka, T. and Yoshimoto, T. (1995) Behavioural response to humans in individually handled weanling pigs. *Applied Animal Behaviour Science* 42, 249–259.

Tennessen, T., Price, M.A. and Berg, R.T. (1984) Comparative responses of bulls and steers to transportation. *Canadian Journal of Animal Science* 64, 333–338.

Terlouw, E.M.C., Lawrence, A.B. and Illuis, A.W. (1991) Influences of feeding level and physical restriction on development of stereotypies in sows. *Animal Behaviour* 42, 981.

Tett, R.P., Jackson, D.N. and Rothstein, M. (1991) Personality measures as predictors of job performance: a meta-analytic review. *Personnel Psychology* 44, 703–740.

Teutsch, G.M. (1987) Intensive farm animal management seen from an ethical standpoint. In: von Loeper, E., Martin, G., Muller, J., Nabholz, A. and van Putten, G. (eds) *Ethical, Ethological and Legal Aspects of Intensive Farm Animal Management. Tierhaltung* 18, 9–40.

Thompson, C.I. (1976) Growth in the Hubbard broiler: increase in size following early handling. *Developmental Psychobiology* 9, 459–464.

Toates, F.M. (1980) *Animal Behaviour – a Systems Approach*. John Wiley & Sons, Chichester.

Turk, D.C. and Salovey, P. (1985a) Cognitive structures, cognitive processes, and cognitive–behavior modification: I. Client issues. *Cognitive Therapy and Research* 9(1), 1–17.

Turk, D.C. and Salovey, P. (1985b) Cognitive structures, cognitive processes, and cognitive–behavior modification: II. Judgements and inferences of the clinician. *Cognitive Therapy and Research* 9(1), 19–33.

van Putten, G. (1969) An investigation of tail-biting among fattening pigs. *British Veterinary Journal* 125, 511–517.

Voith, V.L. (1985) Attachment of people to companion animals. *Veterinary Clinics of North America* 15(2), 289–295.

Walker, S. (1987) *Animal Learning: an Introduction*. Routledge and Kegan, London.

Waring, G.H. (1983) *Horse Behaviour: the Behavioural Traits and Adaptations of Domestic and Wild Horses, Including Ponies*. Noyes Publications, Park Ridge, New Jersey, pp. 235–250.

Weiss, J.M. (1971) Effects of coping behaviour in different warning signal conditions on stress pathology in rats. *Journal of Comparative and Physiological Psychology* 77(1), 1–13.

Index